# GENETIC ENGINEERING

## A Reference Handbook

Other Titles in ABC-CLIO's
## CONTEMPORARY
# WORLD ISSUES
Series

Books in the Contemporary World Issues series address vital issues in today's society such as terrorism, sexual harassment, homelessness, AIDS, gambling, animal rights, and air pollution. Written by professional writers, scholars, and nonacademic experts, these books are authoritative, clearly written, up-to-date, and objective. They provide a good starting point for research by high school and college students, scholars, and general readers, as well as by legislators, businesspeople, activists, and others.

Each book, carefully organized and easy to use, contains an overview of the subject; a detailed chronology; biographical sketches; facts and data and/or documents and other primary-source material; a directory of organizations and agencies; annotated lists of print and nonprint resources; a glossary; and an index.

Readers of books in the Contemporary World Issues series will find the information they need in order to better understand the social, political, environmental, and economic issues facing the world today.

# GENETIC ENGINEERING

## A Reference Handbook

Harry LeVine III, Ph.D.

**CONTEMPORARY
WORLD ISSUES**

**ABC-CLIO**

Santa Barbara, California
Denver, Colorado
Oxford, England

Library of Congress Cataloging-in-Publication Data

Le Vine, Harry.
    Genetic engineering : a reference handbook / Harry Le Vine III.
        p.   cm. — (Contemporary world issues)
    Includes bibliographical references and index.
    ISBN 0-87436-962-2 (alk. paper)
    1. Genetic engineering—Handbooks, manuals, etc.   I. Title.
II. Series.
    TP248.6.L4    1999
    660.6'5—dc21                                          99-28264
                                                            CIP


03 02 01 00 99   10 9 8 7 6 5 4 3 2 1

ABC-CLIO, Inc.
130 Cremona Drive, P.O. Box 1911
Santa Barbara, California 93116–1911

This book is printed on acid-free paper ∞ .

Manufactured in the United States of America

*To my daughter Julia,*
*who along with her generation*
*will live the wisdom or folly*
*of how society addresses*
*the issues raised in this book.*

# Contents

# List of Tables and Figures

**Tables**

# Preface

As nomadic peoples settled down, they brought useful plants under cultivation and gathered herds of animals. They learned to interbreed varieties to get larger, faster-growing stocks. These traditional means of genetic improvement culminated in the Green Revolution of the 1960s, in which high-yield and disease-resistant varieties of plants seemed capable of eliminating the specter of famine from much of the world.

In the early and middle 1970s, yet another revolution was brewing, one that had the potential to take genetic manipulation beyond people's wildest dreams. Human-mediated rearrangement of isolated pieces of genetic material comprising genes from different species, dubbed "recombinant DNA," was achieved in university laboratories. The awesome (or to some, awful) potential power of this genetic engineering caused scientists, in the fall of 1975 at Asilomar, California, to consider for the first time a self-imposed moratorium on certain types of experiments until the risks of the new technology were better understood.

The furor over regulation and application of genetic engineering continues today. Along with uncertainty over effects on the environment, application of genetic engineering emphasizes humankind's demonstrated lackluster competence in addressing social issues posed by new technologies, particularly before a crisis is reached. These range from heartfelt general questions of morality and ethics to privacy issues and access to information or fairness to different groups of people impacted by that information. Despite the often extreme positions taken by the media and various special interest groups, there is room on all of these issues for honest differences of opinion. No single point of view holds all of the "right" answers. The purpose of this book is to provide sufficient background on the issues involved with the application of genetic engineering to allow concerned citizens to participate in the decisions that must be made to fulfill the considerable promise of this new technology while avoiding both ecological disaster and the type of control over human life depicted in 1932 in Aldous Huxley's *Brave New World*.

The first two chapters of this book provide background material for understanding the technical basis of the science and a historical account of the evolution of genetic engineering and the issues surrounding its application. The third chapter sketches the lives of important contributors to the knowledge and the dialog. Chapter four collects data and documents and opinions central to the genetic engineering controversy. Chapters five through seven provide a list of resources for those wishing to delve deeper into the subject, including organizations, print, nonprint, and electronic references. A glossary of genetic engineering terms provides explanations of the technical jargon.

# Overview of Genetic Engineering

# 1

## What Is Genetic Engineering and Why Is It Important Today?

I n the early 1970s, a scientific experiment changed the relationship of humankind to the fundamental processes of nature. For the first time, deoxyribonucleic acid (DNA)—the essence of heredity—was purposefully transferred from one species of organism, *Xenopus laevis,* the African clawed frog, into another species, *Escherichia coli (E. coli)*, the common human intestinal bacterium. Nothing exciting happened. The bacteria grew normally, blithely replicating the piece of foreign DNA from another species that had been inserted into a carrier bacterial DNA plasmid in their cytoplasm.

Although this experiment was in itself only an incremental extension of previous work, the workers in Stanley Cohen's and Herbert Boyer's laboratories had prepared the frog-bacteria plasmid in a test tube using isolated bacterial enzymes to cut and paste the DNA fragments together in a specific order. They had become genetic engineers, rearranging the DNA code for their own purpose. This simple demonstration ushered in the age of genetic engineering. Optimists foresaw that bacteria, yeast, plants, and animals could be modified to produce raw materials for industry, to improve food, to discover new medicines, to remove environmental contaminants, to recycle waste, and to provide permanent cures for inherited diseases.

But there was a cloud over this vision. To some people's minds, genetic engineering changed the world's natural order by mixing genes from one species with those of another species. Ahead lay catastrophic disruption of the earth's ecosystem, uncontrolled spread of microorganism antibiotic resistance with attendant new plagues, and corruption of the ideal of the sanctity of life itself.

## A Brief History of the Genetic Engineering Revolution

Virtually immediately after the groundbreaking experiments there was a call for a moratorium, both from the lay public and from some scientists, to halt certain types of DNA transfer. This was an unprecedented turn of events in the scientific research community. During the summer of 1971, Robert Pollack, a molecular biologist at the Cold Spring Harbor Laboratories on Long Island, New York, convinced Paul Berg and other molecular biologists working on the monkey SV40 tumor virus to put off experiments transferring SV40 DNA that was potentially tumorigenic for humans into *E. coli* cells. They were concerned that since SV40 was capable of infecting human cells as well as monkey cells, incorporating tumor genes into a bacterium that normally colonizes the human intestinal tract could transmit these genes to humans, causing cancer. Berg and Pollack subsequently organized the pivotal Conference on Biohazards in Cancer Research in January, 1973, at Asilomar, California. This conference, called Asilomar I, was to set the tone for recombinant DNA research in the United States and much of the world.

The Asilomar Conference, sponsored by several major government funding agencies including the National Science Foundation, the National Cancer Institute, and the American Cancer Society, was a rude awakening for the scientific community. For the first time, scientific policy was going to be decided not through the traditional peer review process by fellow scientists based on the scientific merit of the research, but by people who had a very different approach, lacked specific training in technical issues, and were pursuing a different agenda. The public feeling was that the potential impact of the new technology on society and the environment was too far-reaching for only scientists to decide what should be done. After all, some reasoned, what had scientists done with the knowledge of how to split the atom? A significant number of scientists were worried also about unintentional transmission of genetic material should recombinant organisms escape into the environment. At the urging of Maxine Singer, a biochemist working on viruses at the National Institutes of Health (NIH) in Bethesda, Maryland, the 1973 Gordon Research Conference on Nucleic Acids held a discussion on the moral and ethical issues of biohazards at the end of the conference. A letter signed by all but 20 of the 142 Gordon conferees was sent to the presidents of the National Academy of Sciences and the Institute of Medicine, and eventually published in the journal *Science*. It urged the establishment of a study committee to recommend specific guidelines "if the Academy and the Institute deem it appropriate."

Reacting to the public concern and the mandate from the scientific community, the National Institutes of Health formed the Recombinant DNA Molecule Program Advisory Committee (RAC) in October, 1974, and charged the group with framing guidelines to govern recombinant DNA research and with reviewing the use of that technology in gene therapy protocols aimed at curing genetic diseases. Further discussions at the Asilomar Conference held in February, 1975, known as Asilomar II, led to a sixteen-month moratorium on recombinant DNA experiments until the NIH Guidelines became available in mid-1976. Ironically, human genetic engineering was specifically excluded from discussion at this time as it was considered too emotionally charged and was too far from realization to be concerned with at that point.

In the meantime, Senator Edward Kennedy and the U.S. Senate Subcommittee on Health of the Committee on Labor and Public Welfare began the first public debate on recombinant DNA in April, 1975. This discussion was echoed at the local level as communities such as Cambridge, Massachusetts, home of Harvard University and the Massachusetts Institute of Technology among other prominent universities involved in DNA research, jointly held town meetings with industrial and university recombinant DNA practitioners seeking to regulate the technology in their jurisdiction.

People were terribly afraid that recombinant bacteria would

escape into their community, causing diseases resistant to modern medicine. Only concerted public relations efforts and information exchanges between the universities and the public averted a fear-driven shutdown of recombinant DNA research at these research institutions. As a result of the early concern, sixteen bills were hurriedly introduced in Congress to regulate recombinant DNA research; none were passed into law. At a National Academy of Sciences forum on industrial applications of recombinant DNA technology held in Washington, D.C., on March 7–9, 1977, dissenters in the audience, in good Vietnam war protest tradition, turned the event from a series of panel discussions into a debate and media event.

Nevertheless, by that time, it was becoming apparent that the doomsday scenarios had been exaggerated and that working with the technology under the 1976 NIH Guidelines was generally safe. In the ensuing years the restrictions of the Guidelines were partially relaxed as safety data accumulated, suggesting that the technology could be controlled. Implications of the application of recombinant DNA technology to genetic engineering emerged as the debate turned from science to social impact.

Many scientists contributed to developing the ideas and methods crucial for making recombinant DNA a useful technology. Some of those contributing the most insightful ideas were honored by the awarding of the Nobel Prize to recognize their accomplishment.

The number of winners whose contributions benefited genetic engineering (Table 1.1) testifies to the development in scientific understanding that was required for the technology to exist.

### TABLE 1.1
#### Nobel Prize Winners Contributing to Genetic Engineering

| Year | Recipient | Field* | Contribution |
|---|---|---|---|
| 1905 | Robert Koch | M&P | Elucidation of pathology of tuberculosis and principles of culture of microorganisms |
| 1910 | Albrecht Kossel | M&P | Studies on chemistry of the cell distinguishing proteins and nucleic acids |
| 1915 | Sir William Henry Bragg | Physics | Analysis of structure by X-ray crystallography |
| 1915 | Sir William Lawrence Bragg | Physics | Analysis of structure by X-ray crystallography |
| 1933 | Thomas Hunt Morgan | M&P | Role of chromosomes in heredity |
| 1958 | George Wells Beadle | M&P | Gene control of cellular chemical synthesis |
| 1958 | Edward Lawrie Tatum | M&P | Gene control of cellular chemical synthesis and genetic recombination |
| 1958 | Joshua Lederberg | M&P | Sexual transfer of genes between bacteria, leading to early genetic engineering |

**TABLE 1.1** *continued*

| 1958 | Frederick Sanger | Chem | Structure of proteins—insulin |
|---|---|---|---|
| 1959 | Severo Ochoa | M&P | Synthesis of RNA and DNA |
| 1959 | Arthur Kornberg | M&P | DNA polymerase and DNA synthesis |
| 1962 | Francis Harry Compton Crick | M&P | Structure of DNA, genetic code |
| 1962 | James Dewey Watson | M&P | Structure of DNA, viral structure, protein biosynthesis |
| 1962 | Maurice Hugh Frederick Wilkins | M&P | Structure of DNA |
| 1965 | Jacques Lucien Monod | M&P | Mechanisms by which genes are regulated and proteins manufactured |
| 1965 | Francois Jacob | M&P | Action of regulator genes, bacterial genetics |
| 1965 | Andre Michael Lwoff | M&P | Replication and genetics of viruses and bacteria |
| 1968 | Robert William Holley | M&P | Structure of nucleic acids, sequence of phenylalanine tRNA |
| 1968 | Har Gobind Khorana | M&P | Synthesis of polynucleotides, the genetic code |
| 1968 | Marshall Warren Nirenberg | M&P | Method for deciphering genetic code, determining protein amino acid sequence from DNA |
| 1969 | Max Delbrück | M&P | Genetics of bacteriophage recombination |
| 1969 | Alfred Day Hershey | M&P | Replication, genetics, and mutation of bacteriophages |
| 1969 | Salvador Edward Luria | M&P | Replication, genetics, and mutation of bacteriophages |
| 1972 | Christian Boehmer Anfinsen | Chem | Control of protein folding by amino acid sequence |
| 1972 | Stanford Moore | Chem | Automatic amino acid analyzer and the sequence of ribonuclease |
| 1972 | William Howard Stein | Chem | Automatic amino acid analyzer and the sequence of ribonuclease |
| 1975 | Howard Martin Temin | M&P | Interaction between tumor viruses and cellular genetic material, and reverse transcriptase |
| 1975 | Renato Dulbecco | M&P | Molecular biology of tumour viruses |
| 1975 | David Baltimore | M&P | Interaction between tumor viruses and cellular genetic material, and reverse transcriptase |
| 1977 | Rosalyn Yalow | M&P | Radioimmunoassay technique |

# The Human Genome Project

Genetic engineering employs the processes used in living cells to reprogram the machinery of other living cells. It seeks to redirect the chemistry in some or all of the many cells making up an organism to change it in some particular way. The changes are made at the level of the genetic information stored in the cell that tells the cell what to do and when, by altering the long strands of DNA code known as the genome. A massive endeavor, the Human Genome Project, is engaged in a 15-year task to determine, by the year 2003, the sequence of the DNA, including 100,000 or so genes, coding for or specifying a human being.

The magnitude of this task can be envisioned by imagining the DNA in the genome stretched end to end, extending around the equator of the earth. One human chromosome would then be 1,000 miles long and one gene would be 1/20 of a mile. Within that gene could lurk a change of one nucleic acid base for another, a mutation, in the span of 1/20 of an inch! Why would one bother to learn sequence information? Because knowing the sequence of the genome is roughly equivalent to having a telephone book with the names and addresses of all of the molecules of the body, as well as a map showing how they are connected.

In October, 1990, the Department of Energy and NIH began the Human Genome Project, eventually forming the National Human Genome Research Institute as the effort grew to determine all 3 billion nucleotides of the DNA in the 23 human chromosomes and to put into place the technologies required to use that information in scientifically, medically, and ethically responsible ways. The Project initially had seven major goals:

1. to map and sequence the human genome with an emphasis on identifying genes;
2. to map and sequence the genomes of five model laboratory organisms—the laboratory mouse *Mus muscalis,* the bacterium *Escherichia coli,* the yeast *Saccharomyces cerevisiae,* the nematode worm *Caenorhabditis elegans,* and the fruit fly *Drosophila melanogaster;*
3. to identify social, legal, and societal issues in order to anticipate and plan for problems that might come to light due to the application of knowledge or technologies arising from the Genome Project (Task Force on Ethical, Legal, and Social Implications, or ELSI);
4. to develop information and analysis systems to allow the Genome Project information to be used worldwide by researchers;
5. to improve technologies for genome study such as DNA sequencing;
6. to facilitate transfer of genome project technology to industry and other areas where it might be useful; and
7. to support training of students and scientists in the different skills needed for genome research.

It was clear that genetic information could potentially be

misused or have unforeseen unpleasant or dangerous consequences. The study of the practical and ethical implications of the
availability and use of human genomic information mandated in
(3) above was included as part of the Genome Project to provide
guidelines for its use and for the protection of individuals (House
Committee on Government Operations, 1992). The potential for
misuse is already receiving the attention of numerous watchdog
genetic resource support groups as well as the federal and state
governments.

The need for full genome sequencing has been hotly debated. Critics oppose the draining of resources away from other
more creative science, questioning the wisdom of determining
the sequence of DNA, 95% of which does not appear to encode a
genetic message, and, finally, ask *whose* DNA would be sequenced. Proponents of genome sequencing reply that the intervening segments of DNA not coding for known products provide
valuable context for the genes themselves that eventually will be
useful. The overwhelming majority of the differences in DNA sequence between individuals reside within the noncoding DNA
sequences. The sequences of genes and the proteins they encode
tend to be highly conserved. Most scientists believe that locating
the human counterpart of a protein whose function has been determined in other animal systems would enormously advance
scientific and medical understanding. Such understanding
would be expected to lead to effective therapies, thus saving lives
and reducing suffering.

While the acute apprehension over handling recombinant
DNA technology eased with the increased knowledge about
safety and with familiarity, concern over the impact of applications of the technology has been building. There remains substantial controversy over the ecological impact of widespread
dissemination of genetically modified organisms. What may
very well turn out to be the real hazard, however, is the social impact of what can be done with genetic engineering technology.
Debate has shifted from the dangers of the technology itself to
what society will do with the information and with the ability to
manipulate genes. The consequences for developing countries of
the economic changes wrought by the new industries born of genetic engineering are yet another area of concern. What was once
a scientific problem now is a social one. How will we use and
control the use of our newfound abilities? It is unsettling to know
that it is entirely in our hands to make the world better—or
worse.

## Basics of Genetic Engineering

### The Structure and Function of DNA

In the late 1950s and through the 1960s, the molecular basis of heredity was elucidated by melding chemical principles with biological observations. As with all good explanations, it turns out that the ideas required to understand the basic principles of molecular biology and genetic engineering are elegantly simple in concept, even though they are not simple to utilize in practice.

The genetic material is made of an unbranched, linear polymer called deoxyribonucleic acid, or DNA for short. The DNA polymer is formed from similar types of units connected end to end like a string of beads, with each bead representing a deoxyribonucleotide unit. DNA is very long and thin—too thin to be seen under a regular light microscope. An electron microscope magnifying nearly a million times is needed to clearly see a DNA strand. Stretched out, the DNA contained in a human cell would be 6 feet long, but it is so narrow that 500 pieces laid side by side could pass through the eye of a sewing needle. (A champion for DNA is the lungfish. Its cells contain DNA that would stretch 1,138 feet, more than 2/10 of a mile!) The DNA is wrapped up in the nucleus, a region of the cell only about a hundred thousandth of an inch across. In order to fit into a nucleus in a skin cell one-hundredth the size of a grain of rice , the DNA is wound tightly to form chromosomes. Each chromosome contains genes aligned along a continuous long piece of DNA. There are 23 pairs of chromosomes in the nucleus of normal human cells. They condense and become visible under a regular light microscope after the cell copies its genome as it prepares to divide into two daughter cells.

The chemical units strung together within a DNA strand are even tinier, so small that the most powerful electron microscope can't make them visible. These bead-like units are nucleotide molecules. Each nucleotide consists of a nitrogenous base attached to a sugar molecule and a phosphate group. The bases come in four varieties, designated by the letters A, C, G, and T, which represent adenine, cytosine, guanine, and thymine, respectively. The backbone of each strand is formed by chemical bonds linking the sugar residues of the nucleotides with the phosphate groups, while the bases protrude sideways. Two single strands of DNA become wound around each other in the helical arrangement illustrated in Figure 1.1.

The bases form the core of the helix. They are held together by weak interactions (dashed lines) like rungs on a ladder with

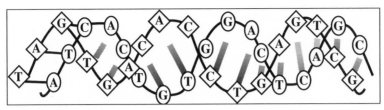

**Figure 1.1. The DNA Helix**

the backbone of the strand facing the outside, similar to the rails of the ladder. The helix can be generated by twisting the ladder lengthwise. It is the sequence of the bases that is important. The letters signifying the order of all of the bases in the DNA of a single human cell would fill a 1,000-volume encyclopedia. Various groupings of these strings are the DNA code of life, signifying different messages. One may tell an enzyme to snip the DNA strands at a certain place while another tells a protein to bind to the DNA somewhere else and yet a third instructs the gene to start production of a protein product needed by the cell.

Having two of these DNA strands wound around each other in the cell guides the copying of DNA when cells divide and helps cells to reduce errors when making new DNA for a daughter cell. Errors, or mutations, could cause problems if they occur in a critical piece of the sequence. New DNA is always copied from one strand of the old DNA. This accounts for how information about hair or eye color or blood type is accurately passed from parent to child. The wound DNA strands are complementary copies of the same code. Where one strand has an A, the other has a T, and where one strand has a C the opposite has a G. A-T's and C-G's pair up, sharing the weak interactions called hydrogen bonds that hold the strands together (see Figure 1.1).

Although at first glance the DNA sequence seems random, in fact the nucleotide units are grouped in "words" whose length depends on the meaning. When DNA directs the production of actual protein products, it uses three-letter "words" to code for amino acids, which are the building blocks of protein molecules. These DNA words are grouped into "sentences" called genes that specify one protein molecule, a polymer of amino acids. Clever experimentation by Francis Crick, Har Gobind Khorana, and Marshall Nirenberg eventually deciphered the code. These special words are first copied in sequence from the DNA into another polymer chain, called messenger RNA, or mRNA (Figure 1.2). RNA is made of ribonucleotide units strung together in a way similar to DNA.

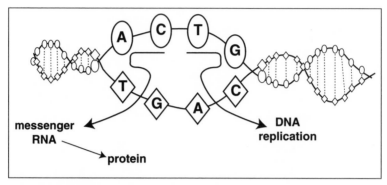

**Figure 1.2. Coding for Protein Synthesis and DNA Replication**

Unlike DNA, many copies of a particular mRNA message are made by the cell. This message is disposable and is destroyed when it isn't needed any more. The message instructs another part of the cell known as a ribosome to join the 20 varieties of amino acid "beads" end to end into polymers to make proteins. It is the order or sequence of the amino acid beads in a protein's polypeptide chain that sets the shape of the protein and determines what it does.

Four, five, six-letter and longer words are read by parts of the cell that control what the DNA sequence is being used for. Some of these longer words specify that certain proteins important for translating the information in the DNA sequence should bind to that site. Examples include the binding of a transcription factor to signal production of messenger RNA leading to protein synthesis, or the binding of DNA polymerase to start DNA replication (Figure 1.3).

There are thousands of ribosomes in a cell, on which many different proteins are made inside the cell at any one time. Proteins

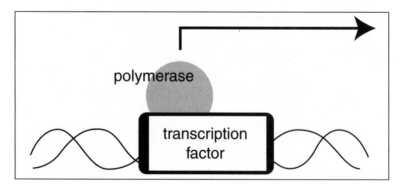

**Figure 1.3. Recognition Sites for Proteins on DNA**

conduct the business of life in a cell. Whereas the DNA is like the card or computer catalog of a library—a list of all of the books in the library—proteins are like the people who use the information in the books from the library.

In multicellular organisms made up of many different types of cells—heart, nerve, and muscle, each cell contains all of the genetic information required to specify a whole organism. Only a small part of the DNA of a cell is being used at any one time, with the particular part depending on the type of cell. The unused DNA is wound up tightly out of the way and twisted onto the scaffold of the chromosome.

Proteins act as catalysts and building blocks to manufacture the other components of a cell, including carbohydrates (sugars) for energy, the cellular protein skeleton, lipids for membranes, and nucleotides for DNA and RNA. Protein catalysts known as enzymes also construct small metabolites from chemicals in their environment. These metabolites are intermediates that transfer needed energy and chemical groups within the cell. Many other proteins help the cell compete for survival or work in concert with its neighbors. In industrial applications with bacteria or fungi some of these metabolites are useful to humans as food (sugars, vinegar), for manufacturing (ethylene glycol, polymers) or as medicines (antibiotics).

Animal, plant, and microbe cells, no matter how complex the organism, use universal code words for the amino acids. Thus, the same protein will be made from the same messenger RNA code in all living things. This feature is what makes biotechnology and genetic engineering work. Whatever type of cell makes a protein from a particular DNA sequence, whether bacterial, fungal, plant, snake, or human, that protein will be the same.

## *Recombinant DNA Technology*

Recombinant DNA technology is the handling and manipulation of the genetic material of cells. The word *recombinant* means "new combinations" and refers to the shifting of genetic material from one organism into another organism of either the same or different species. Nature can also move genetic information from one cell into another by viruses or by "jumping genes," first described by the plant geneticist Barbara McClintock in 1951. Fortunately this happens only rarely under normal circumstances. Scientists called molecular biologists have learned how to speed up and to control the transfer process as well as how to transfer DNA between different species.

Recombinant DNA technology allows the DNA sequence that codes for making a protein (called an insert) to be placed into a small circular piece of DNA called a plasmid vector. A vector is a carrier for the insert DNA which is used to transmit copies into new cells when the host cell divides. Vectors can be designed with control elements to direct the synthesis of mRNA from the insert for the production of protein. These plasmids were modified by molecular biologists from naturally occurring plasmids found in certain microbes or from viruses. They have evolved to persist in cells and to control their own replication. The process, called cloning, is depicted in Figure 1.4.

The plasmid and insert are separately treated with the same DNA cutting enzyme, a restriction endonuclease such as BamHI that generates complementary or "sticky" ends by clipping the backbone connecting the nucleic acid bases at specific sites (see arrowhead). The ends of the strands are labeled 5' and 3' (read "5-prime" and "3-prime"), with the message being read along the sequence in the 5' to 3' direction (Figure 1.4).

The endonuclease specifically recognizes the nucleotide base sequence GGATCC and cuts between the G's in the GG pair at the 5' end of each strand. There are hundreds of different kinds of restriction endonuclease enzymes isolated from different species of bacteria, each recognizing different nucleic acid sequences. The treated insert and vector are mixed and reconnected by the action of the DNA-joining enzyme, DNA ligase, which couples the backbones of the DNA strands. A cell—either a microbial cell or a cell from a multicellular organism—is then "transformed"; that is, it is made to take in the vector containing the DNA insert that codes for the protein. The vector plasmid with its cloned gene is copied when the cell multiplies so that every cell has at least one copy, usually more, of the information. The vector contains DNA sequences that direct the host cell's machinery to make mRNA from the insert sequence, which is then translated into the cloned protein in the cell.

This technique is routinely used to produce a protein in a laboratory for study or in vats for industrial production. It can also be used to provide gene therapy to replace a deficient gene product in cells of a seriously ill patient. W. French Anderson and his fellow medical scientists used just such a trick in 1990, inserting the gene for adenosine deaminase enzyme (ADA) to repair the white blood cells of two children who were always getting infections because their white blood cells could not make this important ADA protein.

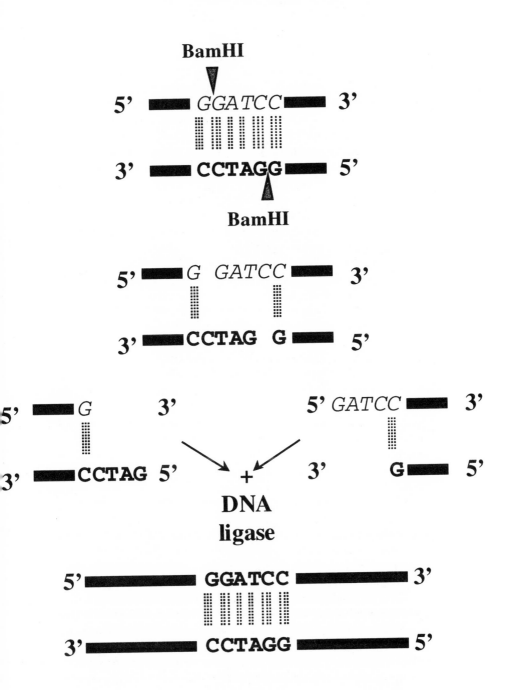

Figure 1.4. DNA Cloning with a Restriction Endonuclease

Recombinant DNA technology is a variety of enzymatic and chemical procedures used to manipulate DNA in the test tube (*in vitro*), to form selected combinations of nucleic acid sequences. By these techniques genes can be added to or removed from the genome of a cell, or existing genes can be modified. The process of changing the genetic complement of a cell or a whole organism by this process has come to be known as genetic engineering. The process of rearranging DNA sequences *in vitro* is key in this definition, because accomplishing similar rearrangements by traditional breeding practices is associated with different regulatory guidelines and a different public reaction. The technology is powerful, permitting much more far-reaching human control over the biological world. It is also more precise, capable of controlling single genes.

Since the advent of the recombinant techniques in the early 1970s, ever more applications of genetic control have been demonstrated in the areas of health, agriculture, industrial production, and environmental remediation. It has also opened up new ways of doing things biologically that had previously been done chemically, if they could be done at all.

Many kinds of cells from different organisms can be used to make useful recombinant proteins from DNA sequences encoded in vectors. Very often these hosts are simple bacteria or fungi that can be grown easily in thousands of liters of culture media. Not only single-celled organisms but plants or animals made up of many cells, such as soybeans, mice, and cows, can also be caused to make a chosen protein. They are known as transgenic organisms, because they carry a gene from another kind of organism, a *trans*gene.

Soybeans harboring a gene making them resistant to weed killers or producing a protein toxic only to insects can make farming them easier. Expressing an altered human gene for a defective protein superoxide dismutase found in some people afflicted with Lou Gehrig's Disease (amyelotrophic lateral sclerosis, or ALS) in a mouse produces some of the symptoms of the disorder in that mouse. This can be used to help researchers find a cure for that neurological disease. Transgenic animals can produce useful human biological products (Velander et al., 1997). The milk from transgenic cows in which particular cow protein genes are replaced with the human protein sequence is a valuable product, an infant formula for babies allergic to cows' milk. These are just a few examples of the usefulness of recombinant DNA technology.

On the other hand, the same technology can pose dangers

through unanticipated effects of a transgene, such as the herbicide-resistant soybean becoming a weed that would be hard to kill or by passing on the resistance to a related plant that could then become a pest. This would be analogous to the antibiotic resistance seen with some bacteria in hospitals before recombinant DNA technology existed. Some people fear the misuse of the technology to create extra-lethal biological weapons or to produce a super race of humans. Others are more concerned with the social and ethical dilemmas stemming from the use of genetic information available through recombinant DNA technology.

An all too easily imagined scenario less fraught with technical details than transgenic weeds is the anticipated threat to personal privacy and possible genetic discrimination resulting from increased genetic testing for legitimate but unrelated reasons. Altering genetic combinations to forms not observed in nature crosses the theological boundary between the powers of God and humankind for some traditional and conservative religious believers. They oppose human interference with parts of nature that they feel that God alone should control.

# Genetic Engineering and Health— A Revolution in Medicine

The unique contribution of genetic engineering technology is the ability to alter specific portions of the genetic code of an organism to repair genetic defects in an individual, potentially curing genetic diseases. Until now people had to accept the genetic background they inherited. Soon, in theory at least, this may no longer be the case. Related DNA technology makes testing for mutations rapid and accurate.

Some attempts at gene therapy have reached the stage of clinical trials. From this experience it is apparent that the benefits of detection and correction of aberrant DNA will come to fruition only after much complex experimental and clinical investigation. The immediate impact of the human genome sequence information will be from the knowledge of the detail of the DNA sequences of genes. This knowledge is already stimulating fundamental and applied research.

Recombinant DNA technology provides tools for understanding better how the body functions in health and disease and for discovering new treatments. This process is explained in a recent *Scientific American* article (Haseltine, 1997). Cellular and animal models that mimic human and animal diseases can be engineered for the testing of new medicines. Health care is only one of several sectors impacted by recombinant DNA technology. Subsequent sections of this chapter will deal with the application of genetic engineering to providing raw materials for industry, enhancing food production, and removing hazardous waste from the environment. Associated problems and potential pitfalls of these applications for society will also be considered.

Genetic engineering, like other technologies such as microelectronics, has the potential to revolutionize the way things are done for the betterment of society, but there are costs to be borne as well. Along with an improved standard of living for many, microelectronics ushered in automation that gradually marginalized unskilled labor with serious economic and societal consequences as low skilled jobs disappeared. Significant changes in society would also be expected with recombinant DNA technology.

Some people assert that difficulties arising from the new genetic technology can be anticipated and controlled. Others fear that dangers of unrecognized proportions loom for genetic engineering. Potential scenarios range from disruption of the earth's ecosphere, with consequent crop failures and widespread starvation, to the development of devastating "germ warfare" weapons.

Less biologically devastating but potentially more socially disruptive would be the use of genetic information to discriminate against people in obtaining insurance coverage or employment on the basis of their genetic heritage. Such applications are a distinct possibility unless controlled by legislation. Ensuring protection of the privacy of an individual's DNA and the ethical implications of choosing the genes of future generations with some gene therapies are also emerging as significant concerns.

## Safety

Many of the early genetic engineering scientists worried about the escape of modified organisms into the environment, their effects on people, and possibly upsetting the global ecology. Their apprehensions were shared by a significant number of nonscientists and public officials who were concerned for public health and envisioned creation of new uncontrollable forms of life. A general lack of understanding of just what could and couldn't be done with genetic engineering was the cause of much unrest. Federal guidelines, first issued in 1976 and updated at intervals, have regulated the types of genetic engineering experiments that can be done. While public debate has largely subsided over the years of apparently safe application of the technology, watchdog groups, like the activist Jeremy Rifkin's Foundation on Economic Trends or the Pure Food Campaign (website http://www.interactivism.com), continue to monitor and challenge the release of genetically engineered organisms into the environment and the marketing of food products from genetically engineered organisms.

## Biological Warfare

In 1969, President Nixon signed a decree unilaterally renouncing biological weapons, purging them from the U.S. arsenal. The Second Asilomar Conference on Recombinant DNA in 1975 was very concerned with the potential of genetic engineering to produce biological weapons of mass destruction. They concluded that ". . . construction of genetically altered organisms for any military purpose should be expressly prohibited by international treaty . . ." In 1975 most countries in the world signed just such a treaty banning production, possession, and stockpiling of biological weapons and toxins. Oddly, the treaty did not apply to chemical weapons (McDermott, 1987).

Although genetic engineering would appear to be ideally suited for creation of devastating disease weapons, many scientists doubt the possibility of intentionally developing a genetically engineered Doomsday Microbe or an Andromeda Strain virus because of a lack of understanding about the causes and limits of disease. Among developed nations at least, the military sector finds such weapons to be too difficult to control to be useful. Since 1975 most nations of the world have renounced the production, storage, or use of biological weapons. Nevertheless, the possibility remains that militarily weak nations might use biological weapons as an inexpensive-to-produce and easy-to-hide "poor man's atomic bomb" or that reckless individuals might release them suicidally without regard to the consequences.

There was concern during the Iraq conflict in 1991 following the invasion of Kuwait that in desperation Iraq would use biological and chemical weapons. The U.S. Department of Defense has maintained a research effort reportedly to provide defense against biological weapons that might be used against the United States. Periodically there are questions about whether the research is straying into the arena of offensive uses, but the issue is so politically charged that the military is very careful about even appearing to be involved with genetic engineering. Public suspicion on this point, though, is fueled by intermittent revelations. In 1992 the Yeltsin government admitted that the Soviet Union had maintained a "standby" program in biological weapons in the 1970s. An apparent leak of four strains of deadly anthrax from Military Compound No. 19 in April 1979 into the city of Sverdlovsk, Siberia, was the first proven escape of an infectious agent from a military biological laboratory that resulted in fatalities. Recombinant DNA technology ultimately provided the proof of this event through polymerase chain reaction (PCRPCR) analysis of frozen autopsy tissue from victims (Hoffman, 1998). Biological warfare remains a highly emotional issue in 1998, particularly with Iraq's periodic refusal to allow United Nations inspections for "weapons of mass destruction" under the peace treaty that ended the Gulf War.

# Genetic Screening

Genetic engineering technology can be used to probe the heredity of individuals. That genetic knowledge carries both health-related and societal consequences. Genetic testing is normally performed on three groups of individuals, the first of which is unborn children (prenatal testing) to determine whether they have inherited a fatal or severely debilitating disease. Adults who want to know whether they inherited a particular genetic trait or a susceptibility for a disease that occurs late in life or whether they could pass a genetic disease on to their offspring are a second population. Lastly, asymptomatic children are tested in a few cases for certain later-developing diseases, but such testing is generally discouraged until the age of majority when the young adult can decide for her-/himself whether to be tested.

Genetic analysis for the presence or absence of altered genes only confirms that the altered gene is present in an individual. Except for a relatively small number of disorders in which the gene product is dominant, inevitably expressing the disease although the time of onset is uncertain, only a risk is associated with the presence of the gene. For nondominant, or recessive genes, to what extent and when, if ever, the disease will occur is unpredictable. For multigenic traits such as heart disease, in which more than one gene is involved, the individual genes are often considered risk factors rather than direct causes. Many environmental modifications such as introducing a diet low in cholesterol and saturated fat can intervene to prevent disease from developing by reducing the risk.

Prenatal testing for biochemical or genetic defects by analyzing amniotic fetal cells or placental tissue is routinely used (300,000 cases per year in the United States) where the family history or advanced maternal age (35 years or older) indicates an increased risk of fetal abnormality such as Down's syndrome.

Newborn babies born in U.S. hospitals are presently screened for a limited panel of anomalies. For the present, most tests are based on biochemical testing for the expressed effect of an altered gene rather than direct gene sequencing. This is because it is harder to determine and interpret risk due to the presence of an altered gene than to find the biochemical hallmarks of the disease before it causes significant damage. Eventually, more genetic tests may be used for the same disorders as the biochemical tests because of their higher sensitivity and simplicity. However, interpreting the results will likely remain uncertain.

Genetic counseling accompanying prenatal testing provides advice to the parents on the possible outcomes of the pregnancy and informs them of their alternatives. Predicting the birth of a baby with medical problems may give parents time to arrange for the baby to be born at a hospital offering a higher level of perinatal care. Many of the genetic diseases detected in pregnancy are presently untreatable, leaving as options only termination of the pregnancy or raising a short-lived or handicapped child with resultant heavy burdens on the family and the child.

Some people question whether genetic counseling about risk is understood as a probability or a certainty by the parents and whether counselors unintentionally steer parents toward abortion. Although the counselor is trying only to make information available while not giving advice, it is difficult to ensure complete neutrality. Other people believe that counselors *should* provide guidance. They reason that if society becomes responsible for the afflicted individual, then society should set some standards for quality of life.

*After* a person is born what are the implications of a diagnosis of an incurable disease of unpredictable severity? How would that alter a person's life? Huntington's disease (HD), a neurodegenerative disorder, typically becomes apparent in adulthood after the disease has been passed to the next generation. In the absence of a treatment for the disease or accurate prognostication of severity, HD family members often do not want to be tested, preferring the insecurity, and hoping that they will experience only a mild form of the disease (see Chapter 4). They feel that HD parents would have a hard time coping with knowing that they have passed on a serious genetic disease to their child.

# Health Privacy

In the United States, particularly, personal privacy is highly prized and zealously guarded. There will be a greater need for confidentiality protection as more potential disease genes are identified. Although medical records—a personal history of what has already happened—are traditionally considered confidential, some people feel that they are not confidential enough. Is a person's genetic code any more confidential than their fingerprints?

Use of anonymous genetic markers for personal identification is similar to fingerprints but more conclusive. Thus their use would be preferable as long as the markers didn't eventually become related to disease loci. At that point, the information would constitute a medical record, which then would dictate a different level of confidentiality. Ostensibly, information from a soldier's DNA sample can be used only for identifying remains, and police can obtain DNA samples under warrant to exclude suspects in a crime, but what guarantees exist that these will be the only uses of those samples? Only five states, so far, have moved to bar unauthorized access to DNA databases.

## Insurance and Employment

The insurance industry stays in business by spreading out the risk of paying settlements over many policyholders, some with a higher risk for payment than others. Companies contend that genetic testing simply adds more certainty to the standard medical testing and family history commonly used in actuarial analyses to determine that risk. By arguing that genetics are beyond the control of the individual, and that risk is not identical to certainty, lawsuits have forced insurance companies to cover individuals with genetic predisposition to a disease and the payment of medical expenses incurred by a child born with a diagnosed genetic disorder. In response, insurance companies and employers may be taking a closer look at people whose *chosen* lifestyle exacerbates their risk.

Similar concern over the inappropriate use of genetic information exists for employment. Cost- and productivity-conscious employers increasingly use genetic screening to select a workforce that will be likely to cost them less to insure and will miss less work due to illness. In 1992, Wisconsin became the first state to forbid discrimination in employment or insurance coverage based on an individual's genetic readout. A 1996 federal law already bars insurers of group plans from considering genetic predisposition a "preexisting condition" in order to limit or charge more for coverage, unless the condition is clinically evident. Self-insured individual plans, however, are not addressed by the statute. In 1997, spurred by the cloning by nuclear transfer of several mammalian species, the federal government faced a wave of anti-cloning, anti-genetic discrimination/genetic privacy legislation. By early 1998 nearly 250 bills in 44 state legislatures proposed limits on access to genetic information. Twenty- four states restrict what insurers can do with genetic information. As in the federal case, the proposed regulations would govern only group policies, not individual policies.

## DNA and the Law

The use of DNA "fingerprinting" as evidence in court was thrust firmly into the public limelight during the O. J. Simpson trial in 1995. It was in the news again in the 1998 investigation of President Clinton's involvement with a White House intern. The principle of the fingerprinting methodology is that the subtle patterns of slight variations in DNA sequence can be quite unique, with the individuality becoming clearer the more gene sequences are examined (Office of Technology Assessment, 1990). This is analogous to the required number of similarities in a regular fingerprint for matching the suspect to the criminal evidence. The legal system now accepts that when properly calculated the probability of matching crime scene DNA to that from a blood sample can be accurately estimated. Such conclusions take into account the somewhat different prevalences of variants of genetic markers (DNA sequence variations) in different racial and ethnic populations.

Genetic markers called restriction fragment length polymorphisms (RFLPs, pronounced "riff-lips") are short DNA sequences, naturally distributed throughout the human genome, that have sequence variations due to random base changes in different populations. They are revealed as altered fragment sizes when the chromosomal DNA is cut with certain restriction endonuclease enzymes as they generate additional cut sites or remove normal cut sites that these restriction enzymes recognize. Originally described in 1978 by molecular biologists Y. M. Kan and A. M. Dozy as identifying the mutation in sickle cell beta-globin, these anonymous (unknown function) markers have been a mainstay of the demonstration of identity between DNA samples by providing a molecular "fingerprint." More advanced techniques employing the polymerase chain reaction (invented in 1980) to copy small amounts of DNA in a sample allowed the analysis of smaller samples, even those partially degraded at the crime scene.

Forensic DNA analysis was first introduced into evidence in 1986 in a United Kingdom immigration case. Acceptance of the new technique was not immediate. It came only after complicated statistical argument based on population genetics probabilities were established as valid, giving the probability that two DNA genetic marker patterns appear identical simply by chance. The "Frye Test," a 1923 U.S. Court of Appeals ruling for the District of Columbia Circuit, was used to judge when evidence obtained through recombinant DNA technology, like any other type of technical evidence, met the criteria for judicial technical evidence, which included general acceptance in the relevant scientific community. Fingerprint and other forms of scientific analysis now commonly accepted had been formerly required to meet similar standards. As with evidence of any other type, DNA evidence can be and has been excluded in some cases if the evidence was shown to have been collected or analyzed improperly.

Unlike a regular fingerprint, a DNA "fingerprint" contains additional information, none of which is used in the forensic determination of whether the defendant was at the crime scene. As more of the human genome is sequenced, more genes are identified and more genes are "mapped," or located with respect to their function or participation in an observable personal characteristic. An "anonymous" or unknown marker, used today only because of its convenience in identifying individuals, could at some time in the future be linked to a disease gene or to a socially undesirable characteristic such as alcoholism or violent behavior. Thus, a DNA record obtained as part of a criminal investigation is also potentially a medical record and may require the same safeguards with respect to privacy. What are the obligations of the authorities to inform this individual of the results of this (unintentional) genetic test and to restrict access to that information?

The unrelated use of the DNA information deposited in a general database from a criminal investigation has become a concern to a growing number of people. Randomly matching crime scene analyses of DNA markers to a general DNA database, "fishing" for a suspect, is considered by some people to be a search without a warrant, a violation of the civil liberty of the individual.

The FBI launched the Combined DNA Index System (CODIS) as a pilot project in 1991 to make DNA evidence available to state and local police departments to aid their investigators. Forensic Science Service's National DNA Database (Birmingham, U.K.) is the only other operational nationwide DNA database (Edgington and Marshall, 1996). In October 1998 the FBI linked the existing 50 state DNA databases through a high-speed internetlike routing system (Wade, 1998). Although the database is resticted to law enforcement use and carries a $100,000 fine for unauthorized disclosures, some civil libertarians fear that its use could be expanded to include almost anyone. They feel that such a system gives the government unwarranted investigative power over its citizens.

## Gene Therapy

Gene therapy is a medical treatment altering the genetic potential of cells. There are currently two types of possible genetic alterations. In one form, changes are introduced into nonreproductive (somatic) cells, often only certain types of cells. These changes affect only the treated individual patients and cannot be passed on to their offspring. The medical practice of gene therapy is in its infancy. Specific delivery of normal genes to the diseased part of the body, replacement of damaged genes, and turning on and off of specific genes are still beyond reach of the "gene doctors."

Altering genes causing a debilitating illness or deadly disease is supported by many people. On the other hand, the use of gene therapy to prenatally "adjust" other embryo characteristics has engendered much controversy. How many parents would opt for enhancing "good" attributes such as intelligence and physical features if such treatments were available? Gene therapy is also expensive, which would mean that only wealthy or well-insured people would effectively have access to the enhancements.

In the second form, germline gene therapy, the permanent alteration of cells, including the reproductive (germ) cells—sperm and ova, causes the altered trait to be passed on to subsequent generations. These germline changes are currently forbidden by government guidelines and by consensus in the medical community. Besides formidable technical barriers to making permanent genetic changes in reproductive cells, people disagree about whether it is right to make permanent changes. Some people believe that although a patient may give consent to his or her own treatment, it is questionable whether that individual has the right to choose for all succeeding generations who might have to bear the consequences of some side effect of the alteration. Furthermore, society may react with prejudice to people who have undergone germline treatment somewhere in their pedigree, but similar to the prejudice experienced by persons who have undergone treatment for mental illness. Senator Thomas Eagleton, George McGovern's vice presidential running mate in the 1972 U.S. Presidential election, was forced to withdraw from the contest when it was revealed that he had undergone mental testing. Conversely, others believe that correcting a deadly or debilitating disease condition to allow a person and their offspring to live a normal life far outweighs the risks, and that NOT providing relief when it is available is wrong. Religious objections to changing "God's handiwork" are also a consideration for some people.

## Human Cloning

While the debate raged over the safety of recombinant DNA technology at the Asilomar Conference in 1973, the issue of human genetic engineering was purposely omitted from discussion. Since it was so far from the realm of possibility at the time, and realizing that strong opinions were held on both sides of the issue, the organizers, probably wisely, chose to keep this volatile topic from derailing the conference—better to concentrate on issues that could be addressed and have some conclusions reached, although the implications of the new DNA technology for human engineering remained in the back of everyone's mind.

The serious dialogue that society needed to have was simply postponed for 24 years, until the possibility again reared its head in the form of the sheep Dolly, who was created from an ovum, an egg cell, by replacing its nucleus with a nucleus from an udder mammary cell, whose genome expressed only mammary cell genes. Although the cell was previously believed to be irretrievably committed as a mammary cell, it was treated in a prescribed manner in an attempt to reverse the mammary cell programming. The transplanted nucleus became nearly embryonic; when implanted in a surrogate mother, the cell was then capable of dividing to produce cells of all the types needed to form a lamb.

The deprogramming of a differentiated genome had already been accomplished with nonmammals in a number of laboratories, but with mammals it just hadn't worked before. Although the success rate was low—1 out of 277 tries—the impossible had now been done, and soon other nuclear transfers with mammals, including two rhesus monkeys, were reported. Many of these were under way at the same time as the sheep experiments. Thus it is likely that this was a case of the technology maturing to the point where it now is *technically* feasible to do a similar procedure with human ova and nuclei.

While the procedure actually used is nuclear transfer, the media immediately christened the accomplishment "cloning," possibly for its added shock value, even though that word is not mentioned in the original published scientific article describing the procedure (Wilmut, et al., 1997).

Although nuclear transfer does not require the recombinant DNA technology that is the heart of this handbook, it is easy to foresee (or imagine) further developments that would allow certain "repairs" or "enhancements" to be performed on the donor nucleus before transfer, making this truly human genetic engineering. While the nuclear transfer technique was perfected to make available animal clones of superior livestock and to facilitate production of transgenic human proteins in these animals, nuclear transfer cannot compete economically with the embryo transfer method that is already in practice for livestock. In this method, embryos are produced by artificial insemination and are then implanted into surrogate mothers.

## Public Response to Cloning

The fervor of the public response, as well as the flurry of condemnation and legal activity, was reminiscent of that greeting the advent of recombinant DNA. A moratorium on such experiments was immediately issued in the United States, and a raft of federal and state legislation is pending. Position statements from scientific societies and religious groups recommending the banning of human cloning have been issued. The FDA has stepped up to fill a perceived regulatory vacuum by stating that any human cloning will require that agency's approval. Internationally, nineteen members of the Council of Europe signed the European Convention on Human Rights and Biomedicine in January, 1998, part of which bans "any intervention seeking to create human beings genetically identical to another human being, whether living or dead." Germany, which already forbids any human embryo research, and the United Kingdom, which was involved in the sheep cloning and aims to continue with its tradition of protecting the freedom of scientific inquiry, did not sign the protocol.

## Religious Implications of Cloning

Implicit in much of the discussion in this chapter is the concern that the application of genetic technologies be fair in respecting the privacy of the individual and that whatever burdens are imposed fall reasonably equitably on different groups in the population. There is a question in some people's minds as to whether some or even any of the technology *ought* to be applied at all. The general uneasiness felt by some people over technology's ability to interfere with the order of the natural world has been stimulated by the somatic cloning of sheep and other animals. The generally held abhorrence of analogous "cloning" experiments with humans has cut across a wide range of societies, religions, and philosophies.

Much the same sort of agreement surrounded the passing of the United Nations Declaration on Human Rights in 1948 when, according to David Tracy in an essay in *Clones and Clones: Facts and Fantasies about Human Cloning* (Tracy, 1998), "Jews and Christians, Muslims, Buddhists, Hindus, Taoists, Confucianists, and people of several indigenous religions found it possible to agree with the practical list of human rights—but each for their own ethical, metaphysical, or religious reasons."

In the case of cloning it would be difficult for these divergent parties to agree on the details of the concepts. For now, they agree on prohibiting one type of manipulation. The concordance may not extend to other procedures. A case in point is the use of animal tissues, whether genetically modified to improve cross-species tolerance or not, in transplants. Pig hearts and baboon hearts have been transplanted into humans whose own hearts have failed and for whom a human donor was not available. This interspecies transplantation is called xenografting.

The Old Testament Bible (Leviticus 19:19) forbids the cross-breeding of distinct plant or animal kinds, prohibits sexual relations between animals and humans, and proscribes eating an animal before the blood has been drained from it. Some Christians, as well as some Jews and Moslems, argue that this type of statement strictly condemns biological intermingling of humans and animals. On the other hand, since genetic material is not exchanged in xenografting, and since many Christians interpret that human beings have been reconciled to God through the death and resurrection of Jesus Christ, thus superseding the ceremonial regulations of the Old Testament, many feel that these medical procedures are allowable. Some argue that even the observed barrier in Nature to crossbreeding between species, indeed the definition of species, is a human construct and not a natural dividing line. Over the millions of years of evolutionary time these barriers constantly change and therefore appear to exist only for a given time, particularly in related branches of the evolutionary tree, such as mammals that share much of their DNA sequences.

Still others would point out that the order of biological systems has developed, whether by self-organizing principles or through the divine intervention of a purposeful Creator, into a smoothly operating whole that should not be disturbed by "mere" human beings because we cannot know the full extent of the consequences of our interventions. Cross-species transmission and adaptation of diseases is a very real possibility, especially worrisome for agents that we might not know about yet. The comfortable, conservative principle of not rocking the boat has appeared in several contexts in considerations of the applications of genetic engineering.

The Judeo-Christian ethic uniquely values human life. Killing of humans is forbidden except in punishment or in time of war, and strictures are placed on appropriate animal use for human benefit such as for food. Do recipients of xenografts become any less of a person? Since grafting does not carry on into the next generation, concerns over succeeding generations and how they will be viewed by their contemporaries should not arise as they do for germline forms of gene therapy that are currently forbidden in medical practice. Although many people accept the breeding and raising of animals for human food, some people have difficulty accepting such practices for organs or tissues, considering this yet another assault on the rights and feelings of animals.

The cloning of humans also raises religious and moral objections about the dignity of a person—treating every human as an individual. This is a consequence of the widely-held belief that no human being should be treated as an instrument to an end. Cloning an incurably-ill child as a "replacement" or creating a genetic double to serve as a tissue donor or even to gratify the ego of parents who want to create a clone of themselves violates this premise. Illogically, having a child "the old-fashioned way" for the same purposes seems to be less of an issue. Such apparent inconsistencies riddle discussions on ethics and morality in general, but despite them some sort of consensus needs to be reached. A practical view suggests that ethicists and the public need to concentrate on what to do with cloning, rather than trying to simply ban it. Many people feel that if it is possible it probably will be done somewhere eventually and making sure that agreed upon standards are present to govern the application offers the best means of exercising meaningful control. Our past experience banning other societal behaviors has taught us that much.

## Possible Impact of Genetic Engineering

Biological science, particularly that applied to medicine and pharmaceuticals, has been utterly transformed by the new genetic technology. Genetic engineering has been tremendously enabling in understanding fundamental biological processes. Scarce hormones, growth factors, and antibodies can be produced on an industrial scale by direct application of the technology. It has supplied rare reagents as well as developed new and useful cellular and animal models of disease to speed the discovery of novel medicines.

On the other end of the spectrum is the decided lack of progress on gene therapy. Gene therapy is something that recombinant DNA technology can do that other treatments can not—remove the root cause of a disease by replacing a deficient gene that caused pathology. Initial concern over human engineering was defused by a general agreement to ban germline gene therapy for the present.

Despite the approval and implementation of hundreds of somatic gene therapy clinical protocols, convincing evidence of a significant effect remains to be shown. Seasoned practitioners point out that this is exactly what would be expected, especially since these treatments were therapy of last resort for most patients. Clinical medicine has always been a slow, trial-and-error process. Changing the pathophysiology was not going to be as simple as popping a new gene into the diseased tissue. Proponents of gene therapy would now agree that they are still learning how to insert genes and have them expressed in the right place at the right time for the proper length of time. This process has turned out to be much more complicated than was originally anticipated.

Although no worldwide ecological cycles stand in peril from human genetic testing, the developed nations are experiencing considerable upheaval and public debate over its social implications. Those countries without the infrastructure or scientific resources for their own recombinant DNA programs may initially be spared the trauma of adjustment to a wealth of personal genetic information, but they will eventually have to face up to its implications as filtered through their own societal norms and sensibilities. They will have the opportunity to observe the debate and types of solutions attempted by the nations who are facing the dilemmas now, and will be able to fashion their own versions. The technology and personal or public safety are not at issue in this case, but the use of the personal information obtained by individuals, the government, and the business/industrial sector to determine personal opportunities and relationships is at stake.

There is no technological fix, no straightforward way to ensure that people are treated fairly while retaining the advantages of having genetic information available. It is a social problem with greatest impact at the moment on the developed nations who are trying to deal with it. The dilemma has been articulated by Linda Bullard (1987):

> Like nuclear power, genetic engineering is not a neutral technology. It is by its very nature too powerful for our present state of social and scientific development, no matter whose hands are controlling it. Just as we would say, especially after Chernobyl, that a nuclear power plant is just as dangerous in a socialist nation as it is in a capitalist one, so I would say the same thing for genetic engineering. It is *inherently* Eugenic in that it always requires someone to decide what is a good and what is a bad gene.

Much controversy centers around the privacy of a person's genetic information and how that privacy will be maintained in an environment where such information, which is out of the individual's control, could be used in hiring and firing decisions and in determining insurability risks. Ensuring control over access to information stored in databases has been an issue since the advent of the widespread use of computers to store and access that information. Although overall security has been reasonable, so far, there are more than a few public instances in which sensitive information has been obtained despite the safeguards. With the creation of extensive databases and DNA banks being projected, such as DNA databases of convicted criminals and military personnel for rapid identification, better control measures able to handle larger amounts of data stored in more places become imperative. At the same time, just such repositories of genetic information, with personal identifying characteristics removed, are indispensable for research into diseases and fundamental biological processes in the twenty-first century.

Iceland's parliament voted 37 to 20 on December 16, 1998, to give the private Reykjavik-based company deCODE Genetics the right to establish and commercially exploit a national database of records from hospitals, clinics, and individual physicians. The Icelandic gene pool is very homogeneous, greatly simplifying the process of looking for disease genes. Although a patient's consent is not required for inclusion in the database, patients may ask that their records be excluded. Opponents are attempting to overturn the law on the grounds of violation of personal privacy (Enserink, 1999).

A foreseeable consequence of the ability to detect and predict the development of the illness in genetic disorders is questions about how such information should be used as well as to whom the testing should be available. Whether to test or not in the first place, it is generally agreed, depends upon whether there is a high correlation between having the gene and having the illness (as in Huntington's disease, Down's syndrome, and cystic fibrosis), which could be used as a basis for reproductive decisions. It also depends on whether an effective treatment for the disorder is available.

The consequence of the inappropriate use of genetic testing results is the *a priori* limitation of personal opportunity on the basis of projected genetic potential. This last issue particularly concerns people who have been diagnosed with a genetic disorder but who do not yet show clinical signs of impairment. It also impacts their families and care givers. The disastrous confusion of genetic probability with certainty for most genetic diseases makes a difficult and unsettling situation for the individual with a positive diagnosis even worse. A genetic predisposition to a disorder is not the same thing as a clinical disorder in which the person or the physician notices the changes due to the disease. Even though the complex interdependence of factors causes many genetic disorders to develop only partially if at all into clinical disease, the afflicted individual is constantly under the pressure of trying to live a reasonably normal life, all the while looking for the first signs of the disease. Should they tell a potential spouse of their condition and should that person also be tested to avoid a union that might produce children with an aggravated genetic risk? Parents may be tempted to divert educational and other family resources to "healthy" offspring with their presumed better chance to make better use of the opportunity.

Mammalian cloning, that is, nuclear transfer as it currently is practiced, does not make use of the recombinant DNA techniques that mark the genetic engineering discussed in this volume. However, the consequences are similar in some respects and as such deserve some comment here. The term "genetic engineering" originally referred to human engineering, which would include reproducing an organism from a single cell rather than its DNA complement. The organizers of the Asilomar conferences in 1973 and 1975 agreed to exclude the topic of human engineering because of the technical infeasibility of such experiments at the time. They also recognized that the extremely emotionally charged nature of such an issue would quickly dominate the public mind and that no consensus would be possible on matters relating to the more mundane issue of the safety of recombinant DNA technology. Indeed, they were correct. The incredible storm of comment and flurry of legislative activity to ban human cloning by both federally and privately funded groups following the February 1997 announcement of the cloning of the sheep, Dolly, from a differentiated mammary tissue cell in Edinburgh, Scotland, was immediately translated in the public view into human somatic cell cloning. The issues of human cloning using somatic cells are closely allied to those of human germ cell gene therapy that has by consensus of the medical community been banned. Besides any theological sensitivities that might be offended, in either of these cases persons would be created that could, at a future time through no action of their own, be subject to discrimination or ostracism depending on the course of societal values.

## Recommendations for Action

When it comes to human and animal health and the research that leads to better and deeper understanding of human health and disease, there is near consensus that genetic engineering has contributed immensely to furthering knowledge and practical pharmacological treatment. One needs to look no further than the enormous private investment in biotechnology companies and in the corporate acquisition of biotechnology capability. Most of this investment activity is in the pharmaceutical and diagnostic sector. People believe that the technology will produce saleable products. On the down side, the miracle gene therapies promised by the new technology have not as yet lived up to their potential in any meaningful way, outside of some judiciously interpreted proof-of-concept cases. Gene therapy has been limited, so far, by government and ethical agreement, to non-germline alterations where only the treated individual is affected, not their offspring.

The arguably most profound impact of the genetic revolution may well be on the ethics of some of the applications of genetic engineering technology. In theory it is possible to overcome the technical objections to safety, ecological disruption, and economic distortions from unequal access to the technology. The philosophical issues are yet another problem. Some of the ethical questions that are being posed now have been disputed vehemently among intelligent people for centuries, regardless of their religious persuasion, without resolution. Genetic testing raises the question of what (as well as who) the individual is, what rights that person has to knowledge about him-/herself, and what rights others have to that same information. What are the rights of future generations and what responsibility does one bear towards descendants? The United States is moving to legislative restriction of both access and use of genetic information for insurance or employment selection. Other countries feel differently about the relative importance of the individual and society and are less restrictive. Outside of this arena, in the U.S. genetic testing used to determine potential predisposition to major genetic disease is presently regulated by a series of medical guidelines depending on the age of the subject and on the possibility of therapeutic intervention in the disease. This is still a very murky area that will continue to expand as more markers for genetic disease are identified.

Nuclear transplantation, popularly dubbed cloning, of humans has leapt to the fore in the public eye as the primary issue in genetic engineering. Not strictly genetic engineering in the sense of recombinant DNA technology, it is not hard to imagine future applications. Although not yet feasible for humans, a great deal of legislative activity all over the world designed to ban similar experiments attests to how close such an activity approaches what many people feel is the limit—the edge—of their humanity. The challenge will be in setting the boundaries on exactly what is going too far without disrupting the beneficial parts of genetic engineering. Finally reduced to fundamental issues of philosophy that are not amenable to proof and reason, this will be the burning question for the new millennium.

# Genetic Engineering of Food

Humankind has experimented with improving food production and food quality since we first domesticated animals and planted crops. Microorganisms were harnessed to brew alcoholic beverages for home and religious use and to process milk into cheese for unrefrigerated pantries. Selective breeding of plants and animals increased food supplies for a burgeoning population. With the advent of modern fertilizers, herbicides, and pesticides and improved methods of land use and livestock management, the Green Revolution of the 1960s hoped to feed the world.

What has the new technology brought about? Genetic engineering has entered into plant and animal breeding projects because of the speed with which desired changes in traits can be made, condensing generations of breeding and cross-breeding over years and generations into a single transfer of genetic information. There are limitations, however; in plants at the present such transfers are restricted to the manipulation of single or small numbers of linked genes, and thus to fairly simple characteristics. Complex multigenic traits involving large numbers of interacting genes are still out of reach. Some changes can be made by transgenesis that will not occur through traditional breeding. The species barrier to transfer of traits can be circumvented by moving, say, the nitrogen fixation capability from legumes to corn or wheat, reducing fertilizer use for those crops. A host of desirable qualities including resistance to plant diseases, herbicides, various pests, salt and toxic heavy metals, freezing, drought, and flooding are being considered as targets for genetic engineering of plants. A cold-regulation gene switch has been identified that controls plant cell defense responses to low temperatures (Pennisi, 1998). Controlling cold response to increase plant tolerance could extend growing seasons and expand the regions supporting agriculture for some crops. Production can also be increased by improving the efficiency of photosynthesis, implantation of nitrogen fixation genes or colonizing factors to attract nitrogen fixing bacteria, controlling ripening of fruit, extending the growing season, and rounding out the nutritional content of major food and forage crops.

## Health Risks

Genetic engineering of food animals has been less well developed, restricted primarily to providing supplements of native growth hormone to cattle and pigs. Treatment of cattle with growth hormone protein produced by biotechnology increases milk production 10% per day and animals reach market weight sooner with leaner meat. Although the hormone given is chemically identical to the natural hormone, there is much controversy over whether products from treated animals and from regular animals should be labeled differently in case problems develop. The European Union still does not allow sale of products from animals treated with recombinant growth hormone.

In many areas of the world, the only source of high quality protein is fish taken from the ocean, lakes, or rivers. Some scientists feel that increasing the amount of edible fish is not just convenient but necessary as overfishing causes commercial fish catches to decline along with the natural fish populations. In fishery laboratories, sockeye salmon, catfish, trout, striped bass, flounder, tilapia, and other species are being given extra copies of fish growth hormone implanted in their DNA (transgenic fish). These superfish grow more than ten times faster than regular fish, require less food per pound of body weight, and are adapted to be raised in "fish farms." These farming advantages have raised concerns about what might happen to the ecology of natural systems if the faster growing and reproducing transgenic fish escaped into the wild fish population.

Although the use of genetic engineering in certain health-related situations is accepted as a necessity by many people, its application in the production of food is much more controversial. At first glance such an attitude is incongruous with the centuries of widespread plant and animal breeding to increase yield and resistance to disease. On the other hand it is an example of the human reticence to accept unfamiliar technology in everyday situations where traditional solutions are available. This is particularly true when the conception for many people is that there is nothing inherently better about the engineered food products, and that the consumer and environment are put at unknown risk simply for the convenience of factory farm producers and manufacturers of processed foods. Although there is some evidence to support these notions, global conditions are changing such that choice among the options may not be freely available to all societies, particularly developing nations and the Third World.

The worldwide implications for large-scale application of plant and animal genetic engineering will be complex and will pit the needs and aspirations of the Third World against the lessons learned by the industrialized nations as the developing nations struggle for their place in the sun. Suman Sahai, the convener of the Gene Campaign in New Delhi, India, has summarized this perspective (Sahai, 1997): "There is little reason for people in food surplus countries to become excited about the biotechnology route to increase the yield of wheat or potato. But can we in India have the same perception? Is it more unethical to 'interfere in God's work' than to allow death from hunger when it can be prevented?"

An issue is whether food products genetically engineered by recombinant methods should be required to be labeled as such and be processed and sold separately from the same products produced through traditional breeding programs. Public health concerns (side effects or allergies), religious proscription, and personal preference considerations supporting special labeling on one hand are arrayed against stigmatizing biotechnology for no particular reason and supporting protectionist economic tactics by countries that lack a significant biotechnology infrastructure. Little useful information about actual product safety is provided to guide consumer decisions by such labeling of engineered crops. Genetically modified soybeans (Monsanto) and maize (Ciba) are being targeted by the European Union, bringing that organization into conflict with the World Trade Organization, which is responsible for overseeing the General Agreement on Tariffs and Trade (GATT). This agreement states that imports can be banned only on scientific grounds. Jeremy Rifkin's Foundation on Economic Trends by way of the Pure Food Campaign, which maintains a generally critical stance on genetic engineering, feels that labeling of all genetically modified produce should be required, and that any resulting conflict with trade law should automatically be superseded by potential public health concerns that could be associated with the genetic alteration. Traditionally cross-bred crops would not be so regulated, a discrepancy often pointed out by proponents of genetically engineered or enhanced foods who feel that such labeling needlessly stigmatizes their products.

## Environmental Safety

There is considerable disagreement over whether genetically modified or transgenic plants grown on a massive agricultural scale represent the same risk or a greater risk to the environment than those traditionally bred. Few controlled studies on the actual risks have been carried out, and then only under a limited set of environmental conditions. Potential problems are seen for many of the proposed genetic modifications. Acquired insect resistance to engineered pesticides such as *Bacillus thurengensis* toxin (Bt) would negate the benefits of the transgene in crop plants as well as reduce the effectiveness of the Bt spores currently used by organic farmers for insect control. Certain plants that overexpress enzyme inhibitors, such as those containing a protease inhibitor or a starch-utilization inhibitor, show insect resistance. Cross-pollination and transfer of herbicide resistance (glyphosate, bromoxynil) to related weed plants could bring problems in weed control that wouldn't be outweighed by the increased effectiveness of smaller amounts of safer herbicides. Viral-resistant plants (for example, squash from Asgrow Seed Co.) have the potential to interact with native viruses already on plants to cause even greater plant destruction, although the actual extent of pathogen evolution in large scale transgenic crops is unknown. Antisense ribozyme, a form of engineered nucleic acid targeted against rice dwarf virus, is being considered to avoid this problem in this important crop. On the other hand, genetically altered fruit ripening (enhanced ethylene production to signal for ripening [DNA Plant Technology] or cell wall hydrolase antisense DNA to retard the fruit softening process [Calgene]) is not expected to have harmful effects on the environment.

The issue becomes muddied when food production and medical benefit come together. Transgenic plants can be made to produce some of the parts of a vaccine, a preparation used to stimulate the body to make antibodies as a host protection against invaders. These immune-stimulating materials contain proteins and carbohydrates (sugars) that are normally part of invading organisms. These can be produced and then extracted from the plants without the disease organism ever being present. Ideally, vaccination by mouth could be achieved by expressing the vaccine components in foods that can be eaten raw. The Sabin oral polio vaccine is given by mouth now, although this route doesn't always work for all vaccines. Vaccines that are produced in food plants and do not have to be extracted to be effective would be ideal in countries lacking access to medicines or refrigerated storage. Researchers at a number of universities and at Mycogen Corporation (San Diego, California) have altered potatoes, alfalfa sprouts, and tomatoes to produce the antigens (the active portion of a vaccine) for hepatitis B, cholera, and traveler's diarrhea. The human antibody molecules recognizing a vaccine can themselves, once their amino acid sequence is known, be made in plants like other proteins for injection to provide short-lived immediate protection without reaction to the human antibody. An antibody against *Streptococcus mutans* bacteria, which cause tooth decay, has been produced in tobacco, a well-studied plant. Plans are being made for formualting these antibodies into toothpaste to fight cavities (Ma, et al., 1995). A different antibody made in soybeans is being used to target drugs against cancer cells.

Inextricably tangled with the "engineered" versus "natural" controversy are the unknown consequences of the widespread use of genetically homogeneous plant populations, a problem presently encountered with traditionally bred high-yield hybrids. Monocultures displace indigenous species and lower the diversity of the gene pool for those traits that are multigenic or not yet assigned to a particular gene. Such populations are particularly vulnerable should a pest or weed overcome the plant resistance and become established. In some sense the dilemmas of today with genetic engineering recapitulate those presented by the Green Revolution in the 1960s. The desertification and aquifer depletion that have accompanied some high intensity agriculture are seen by some as a prime example of the lack of proper management that should be avoided this time.

## Possible Impact of Genetic Engineering

The ecological consequences of the widespread use of genetically modified organisms in agriculture and in industry remain largely unexplored. No application seems entirely risk-free. Even such programs as bioremediation that are directed to redress environmental damage inflicted by industry and a consumer-driven society are fraught with possible ecologic nightmares. Potential scenarios range from the rampant spread of pest- and weed-resistance and cold-tolerance, massive crop failures and starvation, contamination of food crops with plants engineered to produce industrial chemicals, to destruction of the world's biodiversity and genetic reserves. Intermixing of genetically engineered foods and those derived from traditional breeding methods could expose consumers to unanticipated allergens or even potential toxins. These dire possibilities are not all trumpeted by a zealous and vocal few individuals with a dread of any technology or any change, although there are a number of such groups and individuals in the public eye. Various incarnations of these issues and evidence for a measurable effect can be found in governmental and research reports though, of course, the true extent of their impact is unknown.

Those considering the positive aspects of genetic engineering point to maintaining and increasing food production for the burgeoning world population while decreasing the reliance on chemical herbicides and pesticides as laudable goals, worthy of a certain amount of risk. Engineered drought-, salt-, cold-, and heat-resistant plant varieties, grown on previously unusable land, could be augmented by plants with more efficient photosynthesis and enhanced nutritional content. Fast-growing, feed-efficient, lower-fat food animals with enhanced processing characteristics would increase the amount of high quality protein available.

## Recommendations for Action

The agricultural impact of genetic engineering is likely to be only slightly less revolutionary than in medically related fields. Just as everyone gets sick, everyone needs to eat. Genetic technologies have the potential to further increase food crop production, using less fertilizer and fewer herbicides or pesticides, all on poorer soil with less and poorer quality water during an extended growing season in harsher climates.

Not established yet and greatly feared, though, are the possible ecological consequences of mass culturing of plants that could pass their favorable genetic properties to wild type plants that would then become weeds, disrupting both human applications and world ecology. The Environmental Protection Agency is responsible for regulating the release of transgenic organisms in the United States. There are also significant economic changes that would be incurred as well as social changes in the farming profession.

Particularly strident voices have been raised over the marketing of transgenic crops, especially for human consumption. Vocal groups of consumers typified by the Pure Food Coalition oppose the introduction of transgenic crops or products from plants or animals produced with the assistance of genetic engineering technology, citing potential health hazards from unnatural combinations of gene products. Resistance to transgenic foods and transgenic imports in Europe claims similar roots, but accusations of economic protectionism have also been made there. At this point the perceived danger is a matter of personal preference, with little or no convincing data presented. Legislation on labeling foodstuffs known to be free of transgenic products would avoid stigmatizing biotechnology products for unproven dangers and would resemble similar labeling for "organic" products.

Ecological disaster caused by the escape of transgenic organisms is the most feared potential consequence of exposure of engineered microbial and plant organisms to the environment, and the hardest to provide evidence against. Adequate testing in the ecosystem in which the organism will be used is rarely done because of the expense and time involved. Understanding of large ecosystems is also poorly developed at present. Although no significant escape into the wild population has been observed for transgenic plants, this does not reassure some people. Meanwhile, enhanced risk-versus-benefit analyses are being carried out and releases are restricted while more is being learned about microbial and plant ecology.

# Biomanufacturing

Biotechnology has played a role in industrial production of fermented products—cheese, yogurt, alcoholic beverages, and soy sauce, to name just a few—for centuries. Most of the world's supply of organic chemicals was produced by microorganisms prior to 1920 when Standard Oil of New Jersey began chemically synthesizing isopropyl alcohol (rubbing alcohol) from propylene, a petroleum product. Some fifty years later, the oil price rises of the 1970s and the advent of genetic engineering made biologically derived raw materials economically competitive once again. The controlled breeding of crop plants and animals in pursuit of high yields of particular industrial chemical building blocks, such as plant oils with desired characteristics, has continued unabated. Genetic engineering has allowed for quantum leaps in production efficiency of single products in a very short time compared to the years of cross-breeding required by traditional genetics.

## Industrial Feedstocks

In many cases transgenic plants have been the choice over microbes for new sources of industrial chemicals or feedstocks. Two main reasons explain this. Plants directly access the prime energy source, the sun, and large scale cultivation and processing of plants is already practiced. Genetic engineering has provided a new wrinkle in the ability to increase yields of starting materials and to make available new classes of industrial chemical building blocks. Genes that code for materials previously obtained from petroleum or less efficient plant or animal sources can be inserted into either microorganisms or higher plants to enrich for industrially valuable products. The requirements for industrially useful materials differ significantly from the same materials in food products, especially economically, since they face intense competition from petro-derived products. Useful bioproducts other than food may include various oils and fatty acids that are in demand for use in soaps, detergents, cosmetics, lubricant grease, coatings, plasticizers, drying oils, thermoplastics, and varnishes. Erucic acid produced in rapeseed plants engineered by Calgene is used as a lubricant oil and as a starting material for making nylon 13–13. Formerly synthetic polymers used in containers and as textile fibers, such as the biodegradable polyhydroxybutyrate, are now produced in a number of bacterial fermentation systems and recovered from the harvested organisms. Agracetus (Monsanto) has created "wash-and-wear" copolymers with cellulose by transferring the bacterial genes for the polyesters to cotton plants. Other components of the biomass making up the cell wall materials of plants such as the complex organic polymers comprising lignins, and the sugar polymers forming the various types of cellulose and starch are also useful either as building blocks for traditional chemical processes, as fuels, or as fermentation substrates for microorganisms to produce useful products. The obstacle to routine use of bioprocessing and bioproduction of industrial materials is making them cost-competitive with petroleum-based starting materials.

## Bioconversion

By cloning specific genes or even multicomponent metabolic pathways from esoteric organisms that are not themselves usable into "workhorse" strains of bacteria (*E. coli*), algae, fungi (yeast), or higher plants (tobacco), normal cellular metabolic products can be converted into scarce drugs or precursor chemicals for further industrial use. Genencor International, Inc. (Rochester, New York) has engineered a family of enzymes into a microbial organism to catalyze the multistep synthesis of 2-keto-L-galonic acid from glucose, an essential vitamin C intermediate (De Palma, 1998). Molecules useful as medicines are normally produced by specially adapted organisms in minute amounts under particular conditions. The pathways can be optimized and redesigned to yield the desired product even if this product was only a minor metabolite in the original organism. This is a rapidly expanding facet of genetic engineering that is predominant in the pharmaceutical production of medicines. Traditional chemical industry is also beginning to make more extensive application of genetic engineering technology in place of more expensive chemical synthesis, notably of stereoisomers that require very specific 3-D arrangements in space of the chemical groups in a molecule.

# Renewable Fuels

The production of biogas (50%–80% methane) from fermenting garbage, animal and human sanitary waste, and agricultural waste materials has been practiced on a small scale worldwide. A broad range of organic products can be converted to methane or similar fuels by microbial activity through a very low-tech process. The main difficulty comes in expanding the process cheaply to industrial scale. The 5.6 billion cubic meters of methane (equivalent to 5 million tons of petroleum) produced annually by Getty Synthetic Fuel is only a drop in the U.S. energy bucket. In order to achieve economic sufficiency in fuel production microorganisms are being engineered to use low cost energy sources and to function at elevated temperatures in the presence of high concentrations of metabolites and at low oxygen concentrations.

Bioproduction of the ultimate clean fuel, hydrogen, which yields only water on burning, can be carried out by the green alga *Chlamydomonas* or the blue-green alga *Anabaena cylindrica* with energy from light (Benemann, 1996). A two-stage process using photosynthetic (light-using) bacteria is being tested at an Osaka power plant. Fermentation of organic wastes by bacteria in the dark can yield hydrogen mixed with methane as an adjunct to bioremediation to give a clean-burning fuel.

With all these technologies economics will drive the utility of the process. Present yields of hydrogen are in the range of 10%–20% of input with economic sustainability requiring production in the 60%–80% range. Closer control of fermentation conditions and genetic and metabolic engineering of pathways will be needed to attain the break-even point. Germany and Japan have invested significantly in this technology, while the United States lags significantly, spending about $1 million annually on research.

# Biopulping

Chemical treatment of wood fibers is used to prepare wood pulp to make paper and to whiten it, an expensive and environmentally damaging process. A fungus that grows on wood, *Trichoderma virida*, accomplishes much of the same chemistry, partially breaking down the cellulose into a starting material for paper and producing as a byproduct a crude sugar mixture that can be used to feed microbes to do other jobs. Again the problem rests in the industrial scale-up, in speeding up the process, and in reducing costs.

# Biomining

The natural action of indigenous bioleaching bacteria such as the common *Thiobacillus ferrooxidans* and *Leptospirillum ferrooxidans* solubilizes metals from their ores, where they are generally present as chemical complexes with sulfur and oxygen. These organisms that live in the ore deposits grow best in highly acidic solutions, pH 1.5–2.5 (neutral water has a pH of 7), either using energy from sulfur compounds and atmospheric oxygen or, in the absence of air, converting ferrous ($Fe^{+2}$) iron to ferric ($Fe^{+3}$) iron. Carbon and nitrogen to make cellular substances come from atmospheric carbon dioxide ($CO_2$) and nitrogen ($N_2$), while phosphorus comes from soil mineral phosphate ($PO_4^{-3}$). Similar bacterial action can remove sulfur from coal deposits, converting it to sulfuric acid to yield both a commercially valuable acid and low-sulfur fuel coal to reduce sulfate air pollution. Other bacteria, such as *Thiobacillus thiooxidans*, *T. acidophilus*, and *Acidophilium cryptum* that recycle some of the *T. ferrooxidans* products, often grow in association with *T. ferrooxidans*, making ore decomposition more efficient. Bacteria of the *Sulfobus* genus attack ores that are resistant to *Thiobacillus* action. They thrive at temperatures approaching 80°C and can work on ore deposits *in situ*, without excavating the ore. These organisms replace the need for high temperatures and high pressures in industrial processing plants.

Biomining can process lower grade (lower metal content) ores than is normally commercially feasible although the process is slower. Controlled bioleaching can recover metals from low grade ores or mine wastes (<0.5 % metal content) as well as more enriched sources and at the same time can minimize environmental pollution from naturally occurring leaching. It is already a commercially viable operation; in the United States alone, copper worth $350 million and uranium worth $20 million were recovered by microbial processes in 1985. Many metals, primarily copper and uranium, but also including cobalt, zinc, nickel, gold, and lead, can be obtained in this way. Gorham International, Inc., projects that metals worth $90 billion annually will be biorecovered worldwide by the year 2000.

# Bioceramics

A characteristic of biological systems is their ability to organize physical structures. Besides their cellular materials, they can also organize the deposition of inorganic constituents such as bone (hydroxyapatite) and the remarkably intricate yet strong and resilient exoskeletons of shellfish ($CaCO_3$) and diatoms ($SiO_2$). Control of microscopic structures like these would be a boon for industrial uses of ceramics as lightweight, inexpensive replacements for metallic parts in high temperature or corrosive applications. Medical uses of bone substitutes require specific forms of the mineral to be biocompatible in reconstructive therapies. Biological control of ceramic microstructure is an emerging field (Mann and Ozin, 1996).

# Possible Impact of Genetic Engineering

The large-scale application of industrial bioengineering is likely to have a significant impact on the environment. In comparison to most medical applications of genetic engineering, industrial uses typically involve amounts of materials that are several orders of magnitude larger, sometimes running into the millions of tons. Working on such a scale while maintaining close confinement and monitoring of the stages of production with genetically engineered organisms is difficult. Escape or accidental release into the environment from these larger populations could potentially disrupt the ecosystem. A margin of safety is provided by using strains of organisms containing debilitating genes that would put them at a disadvantage compared to the organisms that are naturally in the environment if they were to escape culture containment. However, it is often pointed out that such features are not foolproof.

Similar concerns exist both for modified microorganisms grown in huge vats and for modified higher plants growing in fields. The latter situation involves cultivation of thousands of acres of recombinant plants growing in fields next to plants that might cross-breed, moving undesirable characteristics in both directions. The plants could also be infected with viruses or microorganisms that might exchange genetic material, and the plants are exposed to insects that might also disperse genetic material in unimagined ways. On the other hand, the products being expressed for industrial applications generally do not confer a selective advantage, unlike pest- or herbicide-resistance, so there is less weed potential for these systems. Situations such as biomining in which microorganisms are released or selected for directly in the environment clearly have considerable potential for causing ecological problems, although many of these microorganisms are already present and have adapted to living in the deposits, albeit in small numbers.

## Recommendations for Action

Industrial applications of genetically modified organisms include the traditional culture of microorganisms in closed systems such as sealed vats. Other systems are more controversial. Utilizing genetically modified plants to produce industrially useful building blocks such as certain oils or mixed ester fibers grown on a large scale runs into some of the same environmental problems as other large-scale transgenic plant propagation. Yet, reducing the use of fossil fuels for industrial production in favor of a renewable resource fits with the trend toward ecologically responsible technologies and could be a significant contributor in the next century.

Biotechnology itself has become a substantial industry. The spirit of entrepreneurship still runs high in the United States in that most biotech companies are small, with almost 80% having fewer than 50 employees. Many were started at academic institutions and were nurtured in university "incubators" by administrations hungry for new sources of income as federal higher education funding dwindled. While providing a pipeline to convert academic knowledge to practical applications, the rush to obtain patent protection for ideas or products and the direct connection of academic scientists to industry have drastically affected the university-industry relationship. This has raised questions for many people about the so-called disinterested involvement of many academics and the role the government has played in funding the research. In Europe and the Far East such connections between the universities and industry are even more common than in the United States. In any event, the profound change in the university-industry relationship in the United States is one of the unpredicted outcomes of the genetic engineering revolution.

# Bioremediation—Environmental Restoration

## Overview of Air, Water, and Soil Cleanup

Along with their evolutionary tendency to fill every conceivable ecological niche, microorganisms have developed the capacity to utilize many sources of energy and nutrients in order to live and reproduce. This includes many materials that are toxic to animals and most plants, such as a wide variety of petroleum products, heavy metals such as mercury, lead, and uranium, and a wide variety of human manufactured chemicals that are seen in nature rarely, such as dioxin, or never, such as fluorocarbons.

Processing of organic wastes in small septic systems has had a long history. In an early large-scale application, biogas from human waste was used as a fuel, originally to reduce pollution in a leprosarium near Bombay, India, in 1900. The use of biological organisms to remove toxins and wastes from contaminated materials is known as bioremediation. These living creatures can process trace contaminants present at the part-per-million level or lower in water, soil, and air, or destroy millions of gallons of spilled oil. Where do the organisms come from and where do they go? What does genetic engineering have to offer such a talented resource?

### Organic Chemicals

Organisms capable of metabolizing almost any known organic chemical are already present in small numbers in the soil and water, a product of the constant struggle for nutrients and space. Living in intimate contact with their surroundings, they have evolved the metabolic means to deal with toxic materials in their neighborhood. With large numbers of organisms and heavy selection pressure by toxins, microbes that can't deal with environmental insult don't grow or die out, leaving those that can detoxify the toxins to multiply and take their place.

Microbe populations found near natural petroleum deposits are enriched in species that can utilize the hydrocarbons and aromatic chemicals that were created over millions of years by the decay and metamorphosis of prehistoric plant remains, some caused by microorganisms. They treat an oil spill as a bonanza and multiply rapidly using it as a food source that the other bacteria in the soil cannot use. Each type of organism can usually only effectively metabolize two or three compounds, so a diverse native bacterial population is an advantage over inoculation of spills with pure oil-eating cultures. In the Alaskan *Exxon Valdez* oil spill of 1989, an effective cleanup measure was to spray oil-soaked beaches with nitrogen- and phosphorus-rich fertilizer to encourage the growth of the indigenous oil-eating bacteria that converted much of the oil to $CO_2$. Although effective, bioremediation is considerably slower than intensive physical cleaning, depending upon the environmental conditions. Studies have shown that when the oil runs out the bacteria die down to trace levels again (Swannell, Lee, and McDonagh, 1996).

After the 1991 Persian Gulf War some 20 square miles of oil-saturated sand from sabotaged oil wells in Kuwait threatened both water desalinization plants and the coastal wildlife. Cleanup teams enlisted microbial help from the Kuwait Institute for Scientific Research near Kuwait City. Open water spills have not responded as well to bioremediation treatment.

Commercial use is being made of microbes selected from natural environments. At Kelly Air Force Base in San Antonio, Texas, two kinds of bacteria found in the soil of a paint landfill and a junk pile are combined with a fungus to strip paint from airplanes without solvents while another bacterium degrades the paint.

Chlorine-containing complex organic chemicals are a considerable environmental liability in that they are carcinogenic, break down slowly in the environment, and tend to concentrate in fatty tissue and cause reproductive problems in birds and mammals. Genetic engineering is being used to construct organisms with combinations of metabolic pathways that break down such compounds more extensively and efficiently than the parental organisms in the soil.

## Water-soluble Contaminants

Nitrogen- and sulfur-containing complex organic compounds are generated by mining, coal tar– and oil shale–processing, wood-preserving, and pesticide and chemical manufacturing processes. They endanger water supplies because these water-soluble molecules are not retarded by adsorption to soil components and pass rapidly into aquifers that feed human water supplies. Compounds of this type are substrates for a number of species of native soil bacteria (*Pseudomonas, Corynebacterium, Brevibacterium, Bacillus, Nocardia*) that convert them to simple organic acids, ammonia, and sulfate ions, which then harmlessly enter the normal food chain.

## Heavy Metals

Removal of heavy metal contaminants from soil and water is a different kind of problem. Since these materials are toxic in their elemental form they cannot be broken down into innocuous fragments but must instead be concentrated either for reclamation or for proper disposal. Microorganisms play a part in the natural release of toxic metals from ores and mine waste. Plants, in particular the *Brassicaceae* (mustardweed) family, accumulate metals such as nickel, cobalt, copper, zinc, selenium, and lead in the above-ground parts of the plant; in some cases they accumulate as much as 1% of their dry weight. Further improvements could be obtained by using genetic engineering to incorporate bacterial enzymes to convert the metal ions to the uncharged, less soluble, and less toxic elemental metal. Highly efficient metal transporter molecules used by marine phytoplankton to recover trace metals from ocean water could be inserted genetically into land plants to increase metal uptake. Peptides binding toxic metal ions with high affinity could also be engineered into plants to increase their capacity.

## Possible Impact of Genetic Engineering

Destruction and cleanup of toxic materials on-site without creating massive dangerous waste storage areas and using nature's own mechanisms—it sounds too good to be true. And so it is. By its very nature, bioremediation involves the large-scale release of a few types of microorganisms, possibly genetically modified microorganisms or plants, into the environment. Control of this release and concerns about its effects on the local ecology—such as exchange of genetic material with native populations—are issues that have limited and will continue to limit the dissemination of this technology. Regulatory agencies, local governments, and citizens will have to wrestle with conflicting agendas: the need for detoxification of air, water, and soil, the need to protect fragile ecosystems, and public unease with genetic engineering technology. The Environmental Protection Agency has jurisdiction and regulatory control over the release of modified organisms. Many of the same questions that face other uses of biotechnology in the open environment are also germane here. Asphalt-eating bacteria are suspected to cause significant damage to roads, particularly where weather and heavy loads cause cracking, increasing microbial access. What is the possibility that super paint-eating or rubber-degrading strains could undergo mutations allowing them to grow in a normal environment, attacking houses, cars, and machinery?

## Recommendations for Action

Genetic engineering holds forth the possibility of effecting a massive environmental cleanup of the various toxic and noxious byproducts of industry and of our high standard of living. *In situ* bioremediation of toxic waste in land, water, and air by either indigenous organisms or by engineered ones can be effective and relatively inexpensive without needing processing factories or moving large amounts of materials. By their nature, however, these techniques require release of large numbers of organisms into the environment without being able to control their spread or any disturbance of balanced populations of other organisms. They could even get into underground storage tanks or accelerate road destruction. Microbial ecology is even less well understood than terrestrial or aquatic ecology so it would be even more difficult to predict release outcomes than in the agricultural case of a transgenic corn plant containing a herbicide resistance gene. Further study of microbial ecology could provide needed information to allow the design of biological containment systems that could provide for safe release. Increased use of engineered higher plants that would be easier to contain than microbes could provide much needed remediation capabilities. In contrast to many other uses of genetic engineering where "ruthless corporate giants" and "greedy individuals" are perceived to be risking the public health and safety of others for their exclusive benefit, bioremediation has the potential to provide a positive impact for all parties.

# International Impact of Genetic Engineering

## Status of Genetic Engineering Around the World

Because of the technological complexity of genetic engineering there is a tendency to pay attention only to its impact on those nations with the scientific and technological base to support their own programs in that area. The economic interdependence of almost all of the world's political subdivisions, however, links them in a web of supply and demand for raw materials, goods, and services. Where a given country's economic base fits within this structure depends on its ability to contribute to both sides of the supply/demand equilibrium. This, in turn, depends on the natural resources of that country, including the usual land, forests, water, and minerals, as well as the labor pool that extracts those resources and converts them to a usable form. Nations are classified as "developed," "developing," and "underdeveloped" or "emerging," based on the proportion of their economy devoted to the production of goods and services relative to the proportion required to feed the population. The impact of technological change can be quite different depending on that nation's economic stage of development. This paradigm, admittedly Western in nature, presently dominates world economic structure, given the virtual collapse of the socialist approach of the former Soviet Union.

Given the limited investment that developing and underdeveloped nations can make in the educational and technological resources required for the development of substantial biotechnology programs of their own, these countries may be destined to be net consumers of genetic engineering and biotechnology, at least in the near term. A restricted industrial base combined with a historical role in supplying raw or minimally processed materials to the more developed nations further constricts their options.

Although some people feel that these restrictions are artificially, and therefore unfairly, imposed by the "developed nation's club," it is also true that some nations with limited natural resources in the form of raw materials, such as Japan, have shifted their economies to manufacturing and services by making the required investments in education and technology. The situation is different for each nation, and the same options are not available to all because of the natural situation or historical developments from past commitments. In addition, the conflict is often drawn between rich and poor nations, which adds an ethical dimension to the otherwise sterile consideration of economic and technical questions.

A number of international organizations already concerned with problems of economic development have begun considering the effects of biotechnology on the relationships between developed and other nations. These include the World Health Organization (WHO), the International Council of Scientific Unions, the United Nations Educational Scientific and Cultural Organization (UNESCO), the Food and Agriculture Organization, the United

Nations Industrial Development Organization, and the United Nations Development Program. They have the difficult job of attempting to reconcile the divergent interests of competing nations to reduce the economic crises and social disruption fueled by international trading forces. With a proper set of guidelines adhered to by all parties and with negotiations in good faith it should be possible to minimize further dependency of the underdeveloped and developing nations on the developed nations while ensuring supplies of raw materials and reasonable markets for finished products for the manufacturing economies. Whether or not this will come about remains to be seen. Neither the form these guidelines should take nor the forum for producing and enforcing any such agreements has been established. The following section details the issues at the heart of the discussion and the present framework for dealing with the problems.

## Impact on Developing Countries

The Green Revolution sparked by traditional plant breeding and the application of high production principles and methods provides lessons for what might happen with genetic engineering applied to biotechnology. In many places during the 1960s veritable miracles were achieved. The yields of cereal crops in parts of the Third World (today's underdeveloped nations) doubled or tripled between 1964 and 1988 with the introduction of improved plant varieties and cultivation methods. In the Indian Punjab wheat and rice harvests were improved as were sorghum, finger millet, and cassava in rain forest areas. By contrast, in areas relatively untouched by the Green Revolution, such as Africa, poverty increased. In developing and particularly in underdeveloped countries a growing majority of the world's poor depend primarily on wage employment, not on income from crops raised on farm land. Biotechnology, similar to the Green Revolution, has the potential to relieve the plight of poor people by providing greater and more stable employment, better nutrition, and increased small farm income. At the same time, these technologies stand to harm the poor by displacing labor or otherwise improving the competitiveness of low labor large farms to the exclusion of small farms and their workers. A similar scenario exists in developed countries such as the United States, where the family farm is in danger of extinction in some areas. The displaced workers in the United States, however, have considerably better economic fallback options than those in the Third World.

One of the criticisms of the use of present biotechnology applications in developing or underdeveloped nations is they are directed solely at the marketing demands of developed nations. These include obtaining staple crops with improved health benefits, for instance by lowering the erucic acid content. In the Philippines, yellow maize varieties were improved to feed chickens; unfortunately, at the same time, the amount of white maize, a staple food of the poor people, decreased.

The great majority of the world's undernourished require more calories; only a minority require more protein. Shifting land from staple food to cattle raising or cattle feed production will reduce the calories available. Both for technical reasons and because the genetic engineering capacity is concentrated in the northern, developed countries with different interests, biotechnological and even traditional plant breeding improvements have been slow in coming to tropical staple plant crops such as cereals. Propagation of valuable plants through tissue culture of plant clones, bypassing genetic engineering per se, has been widely applied instead to coconut, palm, potatoes, coffee, and tea. This is the type of biotechnology research most undertaken in developing countries.

A concern of developing and underdeveloped nations is that the biotechnology products developed from genetic engineering represent yet another form of economic enslavement to the developed world. With the current generation of genetically engineered products the benefits of the technology remain with the producer. The pest- or herbicide-resistant plants or F1 hybrid seeds have to be purchased fresh each year from the supplier in the developed country. A growing trend for chemical companies to buy up agricultural biotechnology and seed companies is disturbing to many because it gives these companies the ability to guarantee a market for their own herbicides and other products. Major political and economic philosophies divide national interests, and these also take on ethical overtones. The Green Revolution was researched and developed to a large extent by the public sector—in government facilities and through government-sponsored research grants to universities and foundations; this was consistent with the expanded role of government in society at that time. The present political environment—particularly in the United States, where much genetic engineering research is being done—favors the private sector in which funding for research and development is provided by private investors. The legislative environment in the United States at the same time supports the issuance of patents protecting property rights to organisms and processes central to genetic engineering and biotechnology. The combination has resulted in the escalation of private, for-profit, competitive industry that is more concerned with short-term economic survival than is the public sector, which in theory can take a longer view of issues. A number of factors converge to produce a monopolistic situation with all of the key supplies in the rich world. Only a few of the underdeveloped and developing countries—Brazil, China, India, Mexico, Pakistan, and some others—possess the resources, scale, capacity, or stable political environment to mount a significant public sector biotechnology effort on their own.

Another concern of some developing nations as well as environmentalists worldwide is that genetically engineered organisms, in most cases plants, will be released on a large scale into the environment without sufficient relevant testing to ensure that they will not spread to devastate local ecosystems. The anxiety is based on the same questions about safety that have been discussed in a previous section. There is a temptation for companies to initially market or to test crops in countries without stringent regulations in order to speed up product development or to avoid expensive trials. Environmentalists fear that debt-burdened or corrupt governments may succumb to the lure of promised quick and easy profits for their country (or for private pockets) and introduce a technology before it is proven safe. In some cases these are the same governments that allow the use of inexpensive pesticides and herbicides the rest of the world has banned because they are polluting. These governments could be mortgaging their country's future for today's profits and at the same time promoting ecological disaster.

It is in the best interest of developing nations to work out some kind of plan to reduce the impact of genetic engineering on their economic equilibrium with developed nations. They should do so sooner rather than later. Given sufficient economic incentive, biotechnology can produce substitute products for the exports of developing nations, a development that would have catastrophic consequences for their largely undiversified economies. Escagenetics (California), working on a commercial scale to produce vanilla plantlets by a tissue culture method, is bent on capturing the $200 million annual U.S. market presently dominated by Madagascar. Nestle's (Netherlands) and the Kao Corporation (Japan) are employing genetic engineering to make cocoa butter substitutes. Cacao represents the second most important agricultural commodity in the Third World. Africa accounts for nearly 60% of world production of this crop. A German biotech firm is working on a coffee substitute; yet the Third World is responsible for virtually all coffee production. The United States alone imports $10 billion to $50 billion of coffee yearly from such countries as Colombia, Burundi, Uganda, Rwanda, Ethiopia, and Indonesia.

What leverage do developing countries possess to avoid economic indentureship to the rich developed nations? While insisting on the adoption of uniform patent protection for biotechnology products, the developed nations have asserted a claim to free access to the genetic resources of the plants and animals of the developing and undeveloped nations as a world right. By some estimates, Asia, Africa, and Latin America have provided the genetic base for 95.7% of the world's land, food, and crops (Peritore, 1995). In comparison, Europe, North America, and Japan, comprising the developed nations, contributed only 4.3% of the diversity. The developing nations argue that the "wild" germ plasm has been modified by centuries of cultivation, breeding, and modification and that it should not be a costless resource to be exploited by companies for products that are subject to patent and trade secret and then sold back to the countries providing the original resource. A proposal by the United Nations Food and Agriculture Organization to establish world Gene Banking Centers with free access to all countries was rejected by the United States, the United Kingdom, Germany, France, Denmark, Norway, Sweden, Finland, the Netherlands, and New Zealand. The U.S. Gene Bank at Fort Collins, Colorado, although U.S. government property, has refused grants of germ plasm to countries it had political disputes with, such as the former U.S.S.R., Cuba, Afghanistan, Albania, Iran, Libya, and Nicaragua. The Treaty for the Protection of Biodiversity signed by 153 countries at a United Nations Conference on Environment and Development held June 1–12, 1992, in Rio de Janeiro was not signed by the United States until 1993 because it did not adopt intellectual property laws. Instead, it contained language dealing with regulation of biotechnology, equitable sharing of research/development benefits and technology transfer, national registration of biological resources, rights of indigenous peoples to profits from their own plants and knowledge, and prior informed consent to any biotech testing in any country. The World Trade Organization attempted to address the commercial issues with the Trade Related Aspects of Intellectual Property Rights agreement signed in 1995. Another conference in May, 1998, in Bratislava, Slovakia, attempted to consolidate the differences between these two major international conventions that address the two faces of the debate, the ethical side and the commercial side.

So, what help might genetic engineering bring to the poor of underdeveloped countries? Disease-resistant and drought-tolerant nitrogen-fixing cotton, for example, would be an advantage if it increased the output of cash crops or reduced the need for imports without lowering employment. This would need to be applied judiciously to products that are labor-intensive such as certain vegetable oils, but not to tea, coffee, or palm oil, where a price glut would be devastating to the economy. Genetic engineering can reduce the need for expensive or imported fertilizers, a concern that was exacerbated by the Green Revolution. Improvement of the shelf life of products through modulation of ripening would help poor countries utilize a higher proportion of their harvest as they are the most likely to have inadequate marketing capacity or infrastructure for processing.

## Possible Impact of Genetic Engineering

Developed nations seldom consider publicly the economic impact of the widespread use of genetically engineered organisms on nations that would be competitors for raw materials or that would be net importers of the technology. Whether or not such repercussions are severe is not likely to halt the application of genetic engineering given the strength of world economic forces. However, the resultant economic and political dislocation can be moderated by appropriately sensitive policy implementation. It would be in the developed nations' best economic interest to avoid creating a dependency situation with frustrated and desperate governments in countries whose major export is replaced and who can't raise enough food for their people because the arable land is growing export crops. Food crops for the mostly tropical underdeveloped or developing nations are not the types of plants being genetically engineered, for both technical and economic reasons. Requiring producers to buy high-tech seeds every year from the developed countries who then control the price of the product in their market has been a problem with hybrid strains in the past; genetically engineered strains would only make it worse. Enlightened self-interest would seem to dictate some thought as to balancing the economic consequences against the initial gains. Unfortunately, such action is rarely seen in the real world unless an issue is made of it, and the developed nations again seem inclined to pursue their own interests as evidenced by their strict insistence on protecting the full intellectual property rights operant in Western economic systems while disregarding genetic diversity contributed by the developing and undeveloped nations' plant and animal natural resources.

## Recommendations for Action

An issue often not commented on is the economic impact of the new genetic technologies on the developing and underdeveloped nations of the world. Genetic engineering is a highly technical field requiring a substantial investment in both training and technology to be able to do the research and to meet regulatory requirements. It is no accident that the developed nations, plus a few of the developing nations, have cornered the market on medical and particularly agricultural applications. Developed nations become the sole suppliers of patented engineered seeds and plants that normally must be purchased annually using the previous year's harvest of products in demand by the developed nations. Consequences of this include effective national economic servitude and decreases in indigenous food crop production to plant enough imported seed for export of products. In response there are disputes, such as those aired at the 1992 Rio de Janeiro Conference on Biodiversity, over intellectual property rights and the value of the contribution of genetic diversity provided by the undeveloped or underdeveloped nations to the forerunners of the engineered plants. A compromise needs to be reached, since the developed nations could, if sufficiently pressed, eventually produce synthetic or engineered replacements for the major exports of the underdeveloped nations—an economic disaster for suppliers.

One internationally relevant area in which there is little information available, largely as a result of its classified nature, is the military use of genetic engineering to provide offensive weapons. Despite a treaty renouncing such research, it is difficult to ascertain exactly what the situation is since all parties deny being involved in offensive weapons work. The concern over Iraq's capacity to manufacture and deploy biological weapons of mass destruction and the insistence on international inspections belies the professional military assessment that biological weapons are not sufficiently controllable for use in modern warfare and has rekindled public concern.

# Predictions about Genetic Engineering

In keeping with the inexactness of the fine art of prognostication, neither the proponents nor the critics of genetic engineering can claim to have accurately foreseen either the accomplishments or the disasters predicted for this technology. No plagues of genetically engineered organisms have surfaced and super monsters have not appeared. Yet, despite more than twenty years' practice of genetic engineering, the final word is still out. Many of the medical and agricultural applications have just recently advanced from the laboratory into the real world, but it is clear that the whole operation is getting ready to shift into high gear. Particularly with the large-scale introduction of genetically engineered plants into the environment there is concern over the impact on the local ecology. This will be closely monitored, if not by the companies seeking EPA approval, by the opponents of such releases.

How do the scientists and physicians, industry, universities, ecological activists, politicians, and finally the general public view the place of genetic engineering in the world today? What have been the gains and the losses over the nearly thirty years since the recombinant DNA revolution began? What are the hopes and fears for the future now that genetic technology is poised to touch people's lives through its increasing use in providing food and medical treatment? The interconnection and interdependence of these sectors of society have increased greatly over the years. They are no longer isolated areas of concern; changes in one area have repercussions in all of the others.

## Scientists and Physicians

Hopes were initially high for gene therapy to repair major genetic disorders such as muscular dystrophy and cystic fibrosis as well as to provide cures for cancer. Although numerous clinical protocols for cancer and genetic disorders have been approved and carried out, genetic therapies have not been hugely successful and appear to present little competition to conventional medicine at this time. A variety of practical problems have stalled progress although in principle the therapies show the intended effects. Much further development is needed before genetic treatments become part of the physician's armamentarium. The NIH Recombinant Advisory Committee, which is entrusted with approving gene therapy protocols, issued a report in December,

1995, concluding that clinical efficacy had yet to be conclusively demonstrated for any gene therapy protocol despite anecdotal claims. Thus far there has been consensus in the medical community not to engage in germline genetic therapy in which the afflicted individual's progeny will carry the genetic change.

Genetic engineering has been successful in producing bioactive proteins with pharmacological activity, but in this first wave, outside of white blood cell hormones to stimulate the recolonization of immune cells after chemotherapy and radiotherapy in cancer patients, these molecules have not been the "magic bullets" that were envisioned. Many more of these proteins are in the pipeline under development.

On the other hand, genetic engineering has utterly revolutionized certain areas of biological scientific research, particularly in their application to medicine. Its impact goes far beyond the production of medical biologicals (insulin, interleukins, growth hormone, granulocyte stimulating factor, and monoclonal antibodies) that involve the relatively straightforward production of scarce biological proteins. The tools for today's biomedical research wholly embrace recombinant DNA technology. The data flowing from the Human Genome Project and other organism genome sequencing efforts will help to rapidly define targets for new medicines and may eventually change the way medicine will be practiced. Pharmacogenetics will identify groups of patients that are genetically programmed to process medicines differently. This should reduce the incidence of side effects. Ironically, knowing more in greater detail than ever before may lead to a more holistic manipulation of cellular and organismal properties to attain health.

## Industry

The modern biotechnology industry has been the main beneficiary of genetic engineering. Biological technology had been practiced on an industrial scale since the middle 1800s, when it was primarily limited by production engineering capabilities. The introduction of molecular biology avoided this obstacle and placed the limits only on biological feasibility. Before the 1960s the term "genetic engineering" referred to human engineering, which at that time was science fiction. Today, "genetic engineering" and "biotechnology" are used nearly interchangeably, although biotechnology generally retains the broader definition.

The use of genetic engineering of plants and microorganisms

to provide improved, cheaper feedstocks from nonpetroleum-based materials as well as specialty chemicals has been a natural outgrowth of the traditional biotechnology industry. Cultivation of large numbers of genetically modified plants raises some ecological concerns, but since the modified characteristics for industrial applications do not generally favor survival there is less worry about escape into wild populations. Microbial production of useful chemicals including pharmaceuticals has continued over the years with little public notice. Conventional fermentation of genetically engineered microorganisms is more readily confined and escape can be monitored so there is less concern with this application.

Genetically engineered foods, on the other hand, have attracted a great deal of attention and will continue to be a lightning rod for activist concern. Expansion of the cultivation of genetically engineered crops into underdeveloped countries raises questions about inadequate ecological testing and exploitation of the economies of these nations lacking significant capabilities in biotechnology. In this scenario, developed nations would provide the market but would keep underdeveloped nations dependent on the technology. Lack of intellectual property protection in many of the underdeveloped nations is cited as having prevented the transfer of applied genetic engineering technology to those countries. Compromise between the commitment to accessibility and equity in the 1993 Rio Biodiversity Convention and the benefits of the commercial use of genetic resources will not be an easy process.

The genetic engineering industry likely will grow but at a more normal pace without the blockbuster expectations of the early years. In the near future smaller biotechnology service-based companies will be established to support various forms of genetic testing and DNA forensics emanating from advances in medical understanding and as a result of the mountain of information provided by the Human Genome Project.

# Universities

The development of genetic engineering has caused a realignment of the relationship between the research universities and industry. The many individuals and small enterprises involved in genetic engineering have blurred the distinction between inquiry to understand more about how things work (basic science), which is what normally is carried on by research universities, and applying that understanding to create a saleable product (applied science), normally done in industry. Basic science discoveries are now patented and publicized, often before their soundness and utility have been established. Such behavior impedes the free flow of ideas and defeats peer review of research results, a system that maintains high scientific standards. It also interjects the question of conflict of interest between individuals trying to commercialize their findings and the research community. Meanwhile, many people feel that academic researchers whose work is paid for by public funding from the National Institutes of Health (NIH), National Science Foundation (NSF), or other government agencies are becoming tainted by commercialization as universities and individuals try to cash in on the process. The drying up of government funding intensifies the impact of industrial sources of funding focused on short term (one-to-two-year) applied research. There remains the question—should taxpayers be paying to allow a few people to profit from an idea developed at public expense, even if the product is of general benefit to society?

## Ecological Activists

Although the initial worries about the rampant spread of genetically modified organisms from laboratories were unrealized, some ecologically concerned scientists and citizens feel that the potential for disturbing the balance of natural systems is actually greater today than in the 1970s. During the moratorium period in 1975–1977 a series of experiments were carried out to evaluate the risk of exchanging engineered genetic material between organisms of different types using "crippled" strains of host bacteria. The restrictions on the types of genetic experiments that could be performed were gradually relaxed. Small-scale controlled releases of genetically modified bacteria to colonize host plants and pests were carried out to estimate the risk of perturbing the local ecosystem. Modified plants containing biopesticide or other pest-resistance genes and herbicide-resistance were also tested in small field plots under a variety of conditions. Although such experiments provided little evidence that genetic escape occurred or that pests developed resistance, many people felt that more extensive testing would be required to properly evaluate risk. The early tests were often carried out under duress, after litigation by groups such as Jeremy Rifkin's Foundation on Economic Trends. Detailed study of the risk issues was not possible under those conditions. Test plot sites for modified plants now number in the hundreds, but there remain nagging questions about the effects of large-scale exposure of the organisms to the environment, involving millions of acres of monoculture in different environments with different species of plants and pests in contact with the crop. Similar worries exist about the escape from cultivation of fish or marine organisms made transgenic for growth hormones or other traits to enhance growth and how escapees might outcompete the wild populations.

The quietly growing consolidation of agricultural biotechnology under the control of major chemical companies has been noted with concern by Steve A. Edwards, Ph.D., who observed the recurring vertical business integration of chemicals, seeds, genetic engineering, and pharmaceuticals (Edwards, 1997). Twenty percent of the stock of the last major seed company (Pioneer Hi-Bred) has recently been acquired by Dupont. Examples of large chemical companies and their wholly or partly owned subsidiaries include Monsanto (Calgene, Agracetus, Ecogen, DeKalb Genetics, Holden's Foundation Seeds, Corn States International, Monsoy [Brazil]); Asgrow Agronomics, which is soon to merge with the pharmaceutical giant American Home Products; Novartis (Northrup King, S&G Seeds, Ciba Seeds); Rhone-Poulenc; and the Dow Elanco cooperative venture between Dow Chemical Company and the pharmaceutical company Eli Lilly (Mycogen, DNA Plant Technology, Empresa La Moderna [Mexico]). Besides ensuring a market for their herbicide chemicals with resistant plants, the acquired seed companies provide local outlets for products to generally conservative farmers and access to the seed company's germ plasm of multiple crop varieties developed through traditional genetic means.

Widespread use of recombinant microorganisms in bioremediation in which the modified bacteria or fungi are directly released into the environment is severely hampered by the lack of knowledge about microbial ecology. Cleaning up oil spills or contaminated soil and water, biomining metal ores, and oil and natural gas recovery are socially and economically valuable uses of genetic engineering. Unfortunately they are also encumbered by large risk factors because of the extensive exposure of the organisms to the environment.

## Politics

The intrusion of genetic engineering into the political scene has intensified after a hiatus lasting from the middle 1970s. The recent upswing of political interest has been fueled by vocal public demands for safeguards against decisions being made based on genetic information which to many people felt like invasion of privacy and, like gender and race, was beyond their control. Concern about personal privacy and genetic information has resulted in both federal and state legislation in the 1990s designed to restrict access to and use of genetic information in employment and insurance decisions. At the same time the use of genetic information and the establishment of databases for forensic purposes has expanded. The issue of database access will have to be addressed as more and more genes are discovered and cryptic DNA markers lose their anonymity. Guidelines, the rudiments of which are addressed in recently proposed "Genetic Confidentiality" legislation (such as H.R.306, HR.341, HR.1815, S.89, and S.422; 105th Congress) will be required for the appropriate use of genetic information and proper communication of test results, consequences, and health care choices to clients.

The newest issue of mammalian somatic cell nuclear transfer, so-called cloning, has touched even deeper chords of philosophical and religious faith because it addresses the deeply felt if only poorly understood notion of an individual and the relationship of genetics to that view. Although it is too early to tell, the flurry of proposed broad legislation will likely give way to carefully circumscribed restrictions on certain kinds of human cell manipulations while allowing other mutually agreed medically beneficial work to proceed.

Finally, another round of legislative activity can be anticipated in the future when gene therapy eventually comes to fruition as a safe and effective means of treating certain disease states. A series of issues ranging from consent for the procedure to protection against discrimination and ultimately germline genetic modification will need to be considered.

# What Can Be Done?

The recombinant DNA technology that powers genetic engineering has been in use for over twenty years now. During this period some of the science fiction of an unknown technology, of escaping mutants, of creating monsters and plagues, has been laid to rest. We also have recognized the importance of some new issues arising from the application of genetic engineering technology that were either not apparent or not fully appreciated at the outset. Six areas of concern have emerged: environmental safety, economic impact, genetically engineered food, personal privacy, and lastly the ethics of gene therapy and human genetic engineering.

A high priority concern to many is the impact of the intentional large-scale release of genetically modified organisms into the environment. The introduction of new species into an ecosystem has in the past led to a small but biologically significant fraction expanding into unoccupied ecological niches and becoming pests, as exemplified by the omnipresent starlings or northward-spreading African "killer" bees. Recombinant genetically modified organisms generally contain only one or a few foreign genes and thus do not qualify as new species, since they can still interbreed with an unmodified member of their species. A weed or pest potential is present if the modifications confer a selective advantage over the native species. Agricultural companies have invested heavily in the genetic engineering of traits conferring herbicide-, pest-, and stress-resistance into crop species. Limited testing under a small variety of ecological conditions suggests that development of resistance of pests or the spread of the selective advantage of the new genes into related species is low. Opponents of the release of modified organisms draw an analogy to human drug trials in which low incidence side effects show up only after the drug is on the market and many thousands of people use it. A major difference is that it is hard to recall killer bees or herbicide-resistant weeds once they have escaped into the ecosystem. Activists want more fail-safe mechanisms built into the system. Under present guidelines, once a particular genetically modified organism is tested and certified under one set of conditions, say in the U.S. Midwest, it may not have to be certified again to be used in a tropical ecosystem in which competition for its niche is entirely different. Industrial representatives argue that recombinant organisms should not be treated differently than those that are genetically altered "the old-fashioned way" by traditional plant and animal breeding to maximize desirable traits, and that are not subject to the same testing restrictions. Although no optimal solution has been identified that satisfies all concerned parties, a compromise would be to institute some form of mutually acceptable risk/benefit analysis to win government approval to release genetically modified organisms. Since ecosystems do not respect national boundaries, some form of reciprocity and definition of standards will have to apply internationally.

Similar arguments of ecological imbalance are made for slowing the use of bioremediation, particularly with recombinant microorganisms or plants. Disturbing native communities of organisms by enriching for particular species that can degrade toxic chemicals or change the properties of extractable resources such as oil, coal, and minerals is not risk-free. Potential hazards must be balanced with the potential of bioremediation to effectively deal with the widespread chemical pollution of land, water, and air in many areas of the globe. Microbial ecology is even less well understood than general terrestrial or aquatic ecology. Although the type of internationally accepted risk-benefit analysis applied to crop species would also be useful for bioremediation, it may also be possible to apply certain forms of bioremediation in semi-isolated systems in which the large-scale escape of organisms can be controlled. This would begin to address what most people agree is an increasingly ominous pollution problem.

Those countries that primarily provide raw materials to the developed nations fear that the application of genetic engineering technology will have a disproportionate economic impact, particularly in the agricultural arena, in favor of the developed world and at the expense of developing and Third World countries. As with other advanced technologies, the developed nations with the skill, technology, and knowledge base to utilize genetic engineering effectively and safely end up being sole suppliers of the technology and can exert undue influence on the economies of nations that adopt genetically engineered crops. With the growing internationalization of agribusiness it becomes harder to control monopolies as national responsibility is diffused. Likewise, developing countries without the necessary infrastructure may rush their own version of genetically engineered crops into production, with a different and perhaps inadequate emphasis on safety testing. A uniform international code for testing genetically engineered organisms destined for release into the environment needs to be adopted. It would be fair to the needs of developing and developed nations alike, yet provide safety for the global ecology. Proper compensation to developing and Third World countries for contributing genetic diversity that is the starting point for genetic engineering and traditional breeding programs in developed nations is a sore point and will need some resolution. The developed nations could find a more receptive partner in trade as well as help solve a significant human problem by investing in research into genetic engineering of food plants of the developing nations and the Third World. Helping the producer nations to better feed their own populations instead of simply concentrating on crops for export to developed nations would help reduce their historically justified paranoic concern about exploitation.

Genetic engineering of food plants and animals is a controversial issue in developed nations that have sufficient food from traditional sources. The concerns are, it should be noted, quite different in those societies that are unable to feed their growing populations with traditional agriculture and look to the new technology to potentially fill the gap and the empty stomachs. In the United States and in Europe, though, there is a sizable and certainly vocal group of people that wish to avoid technological enhancements of foods. Their motivations are diverse, ranging from ethical concerns over genetic modification or controlled breeding of plants and animals to fears of chemical and biological contamination and potential allergens. There is a significant movement to ban, restrict, or at least label foods produced or containing components produced with the help of these technologies. Labeling of these products is opposed by the food industry and some governments as unnecessary since it is based on an unproven assertion that the products are harmful, and since labeling is a potential way to impose trade barriers. A solution that would allow those who wish to do so to avoid "high-tech" foods would be to develop a standard to label only those foods produced in the *absence* of technologically enhanced ingredients similar to the "organic" designation of foods.

The most dramatic impact of genetic engineering in the developed countries is likely to result from our increased genetic knowledge about individuals and the ability, or perceived ability, to make predictions about their future health and their unborn children's health. The challenge for a democratic society is how to regulate access to this information yet preserve the ability of individuals to use the information to make decisions about their life.

Maintaining the privacy of the individual is highly valued in many Western societies. In the United States this feeling has been elevated to sacred status, as evidenced by the strong opposition to a national identity card periodically proposed to help control various forms of fraud. Such a system is accepted as a natural course of events by the English and many Europeans. The recombinant DNA revolution and the Genome Project with its computer technology now can place the ultimate privacy, an individual's genetic makeup, within full view. Maintaining confidentiality of medical records, particularly when they contain genetic information, is a growing concern that will continue to escalate as more is understood about the implications of particular pieces of genetic information. Forensic DNA evidence and DNA samples need to receive similar protection as the relationships are discovered between the markers used to match DNA samples to individuals and actual genetic traits, and as law enforcement agencies seek to maintain criminal DNA databases. Federal and some state legislation concentrating on control of the use of such information by employers and insurance companies is in various stages of being implemented; this is more likely to be effective than simply controlling access to the information.

The use and interpretation of testing for genetic diseases is a complex matter as it cuts across the issues of privacy and of ethics. Prenatal testing, testing of children, and testing of adults at risk for a genetic disease have different implications depending on the impact on the individual and the possibility for effective treatment. Guidelines for testing have been proposed by organizations such as the Ethical, Legal, and Societal Implications (ELSI) Task Force on Genetic Testing of the Human Genome Project and the National Society of Genetic Counselors (NSGC). Recommendations for genetic counseling standards of care have been made by the National Institute of Medicine, the American College of Obstetricians and Gynecologists (ACOG), and others. This is still fertile ground for discussion amongst bioethicists, theologians, legislators, and concerned citizens.

Genetic engineering as a medical procedure or gene therapy is recognized by many people to have the potential to cure certain genetic maladies, but as yet the hoped for clinically significant success has not materialized. The presumed eventual development of effective somatic gene therapy will then open the question of germline gene therapy that would affect the offspring of the treated individual, presently recognized as off limits by the medical profession. Does the present generation have the right to make a decision on a therapy that will affect future generations physically and perhaps ethically?

The fear lurking in the background in the early days of genetic engineering that the technology would be used to engineer humans was at the time largely dismissed because it seemed so far fetched. In February, 1997, the cloning of the sheep Dolly from an adult differentiated tissue cell implanted in a ewe rekindled the specter of Huxley's *Brave New World* for many people. Suddenly, it was all possible. Federal legislation was immediately drafted to restrict any human cloning, and scientific and medical societies moved to condemn any such activity by their members.

Such legislation must be carefully written to avoid blocking critical research for disease cures while preventing cloning of humans. Will limited human cloning eventually be acceptable? At some point we will need to consider why human cloning is so ethically abhorrent, and just how cloning differs substantially from normal maternal multiple births.

The only real option for ensuring that the uses of genetic engineering are not injurious is to adopt sensible, practical means of managing its application. Historical experience has (hopefully) taught the United States that outright banning of certain behavior such as alcohol production and consumption during the Prohibition Era in the early 20th century is ineffective. Illegality made things worse by forcing those involved underground away from the tempering effect of public scrutiny and criminalizing large numbers of otherwise law-abiding citizens.

There exists such a diversity of opinion over the risks and benefits for different applications of genetic engineering that a worldwide ban would be impractical. For example, in countries with difficulties raising the food needed for their population and with the potential for engineered food crops that could grow in previously unusable soil, one would likely find little sentiment for extensive time-consuming field trials of these crops. Likewise there are few worries about releasing soil microbes to remove the organic carcinogens and toxic heavy metals from the land, water, and air in Eastern European nations of the former U.S.S.R. trying to overcome decades of industrial pollution poisoning their land and their people. Add to this that not all nations have a government that responds to the will of its people, if they were to express it, and that some nations' leaders have a different agenda. An outright ban would be ineffective in preventing the practice of genetic engineering. A determined individual or group could readily find a safe and secret harbor somewhere in the world to work where there would be no controls in place to ensure safety.

Faced with the above-mentioned possibilities, the most effective policy will likely be to regulate the application of genetic engineering with an appropriate nonoppressive program of controls that uphold reasonable safety standards and a common set of ethical values. No one will be totally satisfied, but all should be able to live with it. Easier said than done, of course, such a compromise will also engender continuing discussion about the issues of fine-tuning the controls in accordance with new discoveries and of accumulating safety data. At the same time, a significant proportion of the issues at stake are not scientific or technical but arise from differences in world view, ranging from rabid technophobia to serious disagreements over philosophical and theological matters. These opinions are often strongly held as part of a traditional belief system and are resistant to disputation and resolution.

# Conclusion

New technologies generally give rise to opportunities as well as to questions and concerns. Genetic engineering is no exception. There are clearly contributions to improving the human situation with enhanced medicines and possibly new treatments in store as the Human Genome project nears maturity and the data become available to the research and pharmaceutical sectors. Less clear will be the gains from bioengineered agriculture that promises industrial feedstocks from renewable plant sources and high quality foodstuffs using fewer chemical fertilizers, herbicides, and pesticides and leaner, faster growing animal products. In order to achieve this, ecosystems will be exposed to genetically modified organisms on a massive scale well beyond any testing. Will they be able to co-exist and can an ecosystem recover from infusion of these types of foreign species? Even for desperately needed bioremediation to clean up our environment the same ecological risks exist.

Beyond questions of safety and the environment rise a myriad of social issues that cannot be settled by experiment and observation. Genetic engineering carries a considerable economic impact as well. Countries able to support the necessary research and development will have an advantage over those that do not. Their capacity to economically pressure the situation may create dependency and further separation of the haves and have-nots.

Maintaining the privacy of an individual's genetic information will be a paramount issue in the face of potential use of that information by employers, insurers, and others. The growth of DNA databases will make controlling accessibility more challenging. Tests for genetic diseases will continue to proliferate and genetic counseling will become increasingly important as more treatment options become available, including gene therapies.

In a sense Aldous Huxley's *Brave New World* is already here. The difference is that we have the opportunity and indeed the obligation to see that it is done right this time. That work of fiction warns us of what we must be aware and what we should preserve. Through careful evaluation of the effects of our new-found technology on the environment and on what we believe the essence of humanness to be, we need to make generally acceptable choices. This will require both education and compromise since a number of the issues rest at the border between rationality and faith. Such differences are hard to resolve directly. They are, nonetheless, solvable at some level though probably not entirely to everyone's satisfaction. This will ensure a continuing and, in the end, a healthy debate.

# References

Benemann, J. 1996. "Hydrogen Biotechnology: Progress and Prospects." *Nature Biotechnology* 14: 1101–1103.

Benowitz, S. 1996. "Researchers View Genetic Testing with High Hopes but Caution." *The Scientist* 10(6, March 18): 1, 6, 7.

Billings, P. R. (editor). 1992. *DNA on Trial: Genetic Identification and Criminal Justice*. Plainview, NY: Cold Spring Harbor Laboratory Press.

Bodner, W., and R. McKie. 1995. *The Book of Man: The Human Genome Project and the Quest to Discover Our Genetic Heritage*. New York: Simon and Schuster.

Bullard, L. 1987. "Killing Us Softly: Toward a Feminist Analysis of Genetic Engineering." In *Made to Order: The Myth of Reproductive and Genetic Progress* (Sapllone, P., and L. Steinberg, editors). New York: Pergamon Press.

Edgington, S. M., and A. Marshall. 1996. "FBI Unveils $48 Million Project for Tracking Criminal DNA." *Nature Biotechnology* 14(June): 691.

Edwards, Steve A. 1997. "Agricultural Biotech Industry Riding Wave of Consolidation." *Genetic Engineering News* 17(19): 1.

Enserink, M. 1999. "Iceland OK's Private Health Databank." *Science* 283:13.

Haseltine, W. A. 1997. "Discovering Genes for New Medicines." *Scientific American* (March): 92–97.

Hoffman, I. 1998. "Piece Found in Anthrax Mystery: Lab Sheds Light on Russian Leak." *Albuquerque Journal* (February 3): from website http://www.abqjournal.com/scitech/.

House Committee on Government Operation, U.S. Congress. 1992. *Designing Genetic Information Policy: The Need for an Independent Policy Review of the Ethical, Legal, and Social Implications of the Human Genome Project (16th Report)*. No. 102–478 USGPO. Washington, DC: U.S. Government Printing Office.

Jackson, J. F. 1996. *Genetics and You*. Totowa, NJ: Humana Press.

Kevles, D. J., and L. Hood (editors). 1992. *The Code of Codes: Scientific and Social Issues in the Human Genome Project.* Cambridge, MA: Harvard University Press.

Ma, J. K.-C., et al. 1995. "Generation and Assembly of Secretory Antibodies in Plants." *Science* 268: 716–719.

Mann, S., and G. A. Ozin. 1996. "Synthesis of Inorganic Materials with Complex Form." *Nature* 382: 313–318.

McDermott, J. 1987. *The Killing Winds. The Menace of Biological Warfare.* New York: Arbor House.

Office of Technology Assessment. 1990. *Genetic Witness: Forensic Uses of DNA Tests.* OTA-BA-483, July. Washington, DC: U.S. Government Printing Office.

Pennisi, E. 1998. "Transferred Gene Helps Plants Weather Cold Snaps." *Science* 280: 36.

Peritore, N. P. 1995. "Biotechnology: Political Economy and Environmental Impacts." In *Biotechnology in Latin America: Politics, Impacts, and Risks* (Peritore, N.P., and A. K. Galve-Peritore, editors). Wilmington, DE: SR Books.

Rifkin, J. 1998. *The Biotech Century: Harnessing the Gene and Remaking the World.* New York: Jeremy P. Tarch/Putnam.

Sahai, S. 1997. "Developing Countries Must Balance the Ethics of Biotechnology against the 'Ethics of Poverty.'" *Genetic Engineering News* 17(10): 4.

Swannell, R. P. J., K. Lee, and M. McDonagh. 1996. "Field Evaluations of Marine Oil Spill Bioremediation." *Microbiological Reviews* 60(2): 342–365.

Thompson, L. 1994. *Correcting the Code: Inventing the Genetic Cure for the Human Body.* New York: Simon and Schuster.

Tracy, D. 1998. "Human Cloning and the Public Realm: A Defense of Intuitions of the Good." In *Clones and Clones: Facts and Fantasies About Human Cloning* (Nussbaum, N., and C. R. Sunstein, editors). New York: W. W. Norton.

Velander, W. H., H. Lubon, and W. N. Drohan. 1997. "Transgenic Livestock as Drug Factories." *Scientific American* (January): 70–74.

Wade, N. 1998. "FBI Plans DNA Database to Help Track Criminals." *Ann Arbor News* (October 12): A01.

Weiss, R. 1999. "Seeds of Discord. Monsanto's Gene Police Raise Alarm on Farmers' Rights, Rural Tradition." *Washington Post* (February 3): A01.

Wilmut, I., A. F. Schieke, J. McWhir, A. J. Kind, and K. H. S. Campbell. 1997. "Viable Offspring Derived from Fetal and Adult Mammalian Cells." *Nature* 385: 810–813.

# Chronology of Genetic Engineering 2

Although the high-technology aspects of genetic engineering have come to public attention mainly since the recombinant DNA revolution in the 1970s, similar activities have actually taken place over the thousands of years of human history. For much of that period people were searching for the order in nature to explain how phenomena were related to one another. Only in the latter part of the nineteenth and twentieth centuries did knowledge of genetic processes progress to the cellular and molecular level to the extent that scientists now feel that they understand much of the detail. Using this insight they have begun to manipulate genetic systems in a variety of ways. The following chronology highlights key developments that led to today's understanding of how genetics works and events that were important in society's reactions to science's new capabilities.

ca. 6000 B.C.    The Sumerians and Babylonians in Mesopotamia produce alcoholic beverages, cheese, and other products by action of naturally occurring yeast present in foodstuffs.

ca. 4000   The Egyptians discover how to make leavened bread
B.C.       using yeast.

ca. 1300   Additional fermentation processes are in use in China,
B.C.       including winemaking in the Shang Dynasty.

1790       The United States passes the first patent law to provide
           commercial protection for inventors.

1804       After much discussion and experimentation, chemists
           conclude that substances consist of identical invisible
           particles called molecules that are themselves formed
           from more elementary particles that John Dalton calls
           atoms. An early theory of the atomic nature of matter
           had been stated by the Greek philosopher Democritus
           (ca. 460–352 B.C.).

1810       The chemical equation for alcoholic fermentation
           $[C_6H_{12}O_6 = 2\ CH_3CH_2OH + 2\ CO_2]$ (conversion of sugar
           to alcohol and carbon dioxide) is deduced by French
           chemist Joseph Louis Gay-Lussac, explaining the
           process that has been in use for millennia.

1836       The processes of putrefaction and fermentation are at-
           tributed to microorganisms by German physiologist
           Theodor Schwann. The French chemist and microbiolo-
           gist Louis Pasteur (1822–1895) later provides experi-
           mental evidence for this in the 1860s. Such explanations
           do not conform to popular sentiment.

1852       Sparrows are imported from Germany into the United
           States to deal with caterpillars that are ravaging food
           crops. This example of the spread of a foreign species in
           a new environment is thought by some in the twentieth
           century to show what can happen with a new geneti-
           cally engineered plant or animal carrying foreign genes
           that give it an advantage.

1862       The Organic Act establishes the U.S. Department of
           Agriculture, removing it from the Patent Office and di-
           recting the department to collect and distribute seeds
           and plants to farmers.

1865    In a presentation before the Natural Science Society in Brunn, Austria, an Augustinian monk from Brno, Gregory Mendel (1822–1884), proposes that invisible units called "factors" are passed from one generation to the next and that they account for transmission of observable traits. Ignored at the time in the wake of the furor over Charles Darwin's evolutionary theory introduced in 1859, Mendel's publication is independently rediscovered by botanists Hugo DeVries (France), Erich Von Tschermak-Seysenegg (Austria), and Carl Correns (Germany) in 1900.

            Louis Pasteur develops the germ theory of disease from his work on the silkworm disease plaguing the French silk industry. The English surgeon Joseph Lister begins using disinfectants such as carbolic acid (phenol) during surgery, in the handling of patients, and in the care of wounds, dramatically reducing patient deaths from infection.

1869    Swiss biologist Johann Friedrich Miescher isolates DNA from white blood cells in the pus obtained from discarded bandages. Since he is busy investigating the nature of the many chemicals found in cells, he fails to make the connection between DNA and Mendel's hereditary factors.

1877    Louis Pasteur notes that some bacteria die when cultured with certain other bacteria. He suggests that some bacteria produce substances to kill other bacteria but goes no further with the concept.

1879    German biologist Walter Flemming discovers structures in the cell nucleus that are stained by certain colored dyes. In 1888, another German biologist, Heinrich Wilhelm Gottfried Waldeyer, uses the word *chromosome* to describe these structures. It is 14 years before the connection is made to the chromosomal role as a repository for genetic information.

1882    German bacteriologist Robert Koch first determines the cause of a human microbial disease with the identification of the tuberculosis organism. In the process, he

1882    establishes that specific diseases are caused by specific
cont.   organisms, by isolating pure cultures of bacteria from
        infected individuals. Similar observations and culture
        techniques are used by Pasteur to produce vaccines
        against anthrax and rabies.

1885    Vaccination becomes an acknowledged weapon in the
        medical arsenal when nine-year-old Alsatian Joseph
        Meister is successfully treated for rabies by Louis Pas-
        teur. Although a desensitization treatment has been
        practiced by the Chinese for centuries, and a smallpox
        vaccine has been developed by English physician Ed-
        ward Jenner in 1796, the presence of disease-causing
        microorganisms had not been determined till now.

1889    The vedalia beetle, commonly known as the ladybug, is
        imported from Australia into California to control the
        cotton scale insect devastating the citrus orchards there.

1890    The discovery of antibodies by German bacteriologist
        Emil Adolph von Behring provides part of the explana-
        tion of the protective factors elicited by vaccines. The
        special cells that carry out the rest of the functions nec-
        essary for immune system surveillance of foreign in-
        vaders are described by Russian zoologist and
        bacteriologist Elie Metchnikoff around 1891.

1892    Viruses are described by Russian bacteriologist Dmitri
        Iosifovich Ivanovski as disease-causing agents smaller
        than bacteria.

1893    The German Nobel Prize–winning chemist Wilhelm
        Ostwald proves the catalytic nature of protein compo-
        nents of cells called enzymes that are responsible for
        cellular chemical reactions such as the fermentation of
        glucose. One enzyme molecule causes the conversion
        of many substrate molecules into product. These bio-
        logical catalysts greatly speed up the rate of cellular re-
        actions that would otherwise occur much more slowly
        without them, if at all.

1895    A German company, Hochst am Main, sells commer-
        cially cultured *Rhizobium* isolated from root nodules to

enhance soil nitrogen fixation in Europe. These soil bacteria are introduced into the United States in the following year.

1899    Chemical modification of the salicylic acid extracted from willow bark and other plant extracts used to combat fever yields acetylsalicylic acid, which displays reduced stomach irritation. Bayer, a German chemical company, markets this first "wonder drug" as Aspirin.

1901    The chemist Franz Hofmeister advances the theory that the vital reactions of life are performed by enzymes, those catalytic molecules that make chemical reactions efficient enough to power life processes. Direct proof for this comes when mutations (genetic changes) crippling enzyme activities are shown injurious or lethal for bacteria, fruit flies, and people.

1902    Proteins are shown by German chemists Emil Fischer and Franz Hofmeister to be strings of individual amino acid units chemically attached end-to-end through peptide bonds to give polypeptides. This theme of using the same chemical reaction to link varieties of a common building block into long chains or polymers to obtain desirable properties is repeated in biology with DNA, RNA, and polysaccharides.

Walter Stanborough Sutton, a U.S. cytologist and later physician, suggests that Mendel's "factors" reside on chromosomes and are segregated as the chromosome pairs separate during the formation of gametes through meiosis. He names these factors genes, the term we use today.

1903    The first use of the term *biochemistry* to describe the chemistry of life is attributed to chemist Carl Neuberg.

1905    A system for analyzing the occurrence of observable traits (phenotypes) based on the genetic principles of linkage and gene interaction is used by William Bateson and Reginald Crundall Punnett. These pioneering English geneticists coin such genetic terms as F1 and F2 generations, allelomorphism, homozygote, and

1905
cont.

heterozygote. The molecular events underlying the observations of the inheritance of traits are eventually
worked out over half a century later.

1907

The culture of isolated cells or tissue separate from the
intact organism (*in vitro*) is established by U.S. biologist
Ross Granville Harrison. Refinements in these techniques allow *in vitro* culture of many types of cells, a
boon to the study of cellular biochemistry. The techniques are eventually extended to tissue and cellular
transplants, monoclonal antibody production, plant
propagation, and genetic engineering modification of
intact organisms from the 1970s onward.

1908

The English physician Sir Archibald Edward Garrod
recognizes that the final product of a gene is a protein
and pioneers studies of genetic diseases in humans
with his work on alkaptonuria. His postulate that diseases can be caused by mutant genes goes unnoticed
until it is rediscovered in 1940.

1909

Wilhelm Johannsen, a Danish botanist, first uses the
terms *phenotype* and *genotype* to describe an observable
trait and the genetic factors (genes) responsible for the
trait, respectively. Johannsen also describes the process
of selection for a genetic trait where an advantage con-
ferred by a gene leads to a higher occurrence of that gene
in the next generation because it enhances survival.

1910

Thomas Hunt Morgan, a U.S. Nobel Prize–winning geneticist, establishes the basis of modern genetics by
proving that genes reside on chromosomes, pinpoint-
ing genes to particular chromosomes of the fruit fly,
*Drosophila melanogaster*. Over the next ten years Morgan's group describes gender-linked genes (white eye
color in *Drosophila*) and other trait linkages. Morgan
and colleagues publish the classic text *The Mechanism of
Mendelian Heredity in 1915,* formally setting forth the
principles of gene theory and linkage.

1912

Butanol and acetone are produced industrially via microbial action using a process developed by the Russian-born American chemist Chaim Weizmann, later to

become president of Israel. This is the first large-scale use of microbial processes for products other than food.

1917      A virus that infects and destroys bacteria, a bacteriophage, is independently discovered by English bacteriologist Frederick Twort and French bacteriologist Félix Hubert D'Herelle. Bacteriophage, such as the T4 and the lambda phage, are important in early studies of gene structure by U.S. physicist-turned-geneticist Seymour Benzer and others in the 1960s. Use of the lambda bacteriophage as a gene-cloning vehicle begins in the 1970s.

1926      The first crystallization of an enzyme, urease, by James Batcheller Sumner proves it to be a protein. He is awarded the Nobel Prize in chemistry in 1946 for this achievement.

Henry Agard Wallace founds the Hi-Bred Company, a hybrid corn seed producer, known in the 1990s as Pioneer Hi-Bred International, Inc., a company involved in genetically engineering agriculturally useful plants.

1927      The first localization of a specific gene to a particular chromosome in mammals (mouse) is accomplished by geneticist Theophilius Shickel Painter who correlates a visible chromosomal deficiency (partial deletion) with a genetic diagnosis.

1928      A transforming principle in *Pneumococci* shown to be DNA in 1944, is demonstrated by Frederick Griffith to convert nonvirulent to virulent strains—a "natural" (that is, unassisted by humans) form of gene transfer.

The first clinically useful antibiotic, penicillin, is discovered by English bacteriologist and Nobel Prize–winner Sir Alexander Fleming, being produced by a mold growing on a contaminated bacterial culture plate. Other antibiotics known at that time are toxic to the mammalian host. Penicillin is isolated in 1938 by chemists Howard Florey and Ernst Chain of Oxford University in England. Although industrial production is initially delayed by lack of patent protection, the antibiotic is eventually made available for clinical use some 15 years later.

1932    A primitive microscope using electrons for sample illumination is built by M. Kroll and Ernst August Friedrich Ruska. High energy, and thus short wavelength, electrons allow DNA and single protein molecules to be visualized, objects some 1,000 times smaller than can be seen with a regular light microscope. Being able to see what previously could only be imagined allows scientists to work out how molecules are assembled to form the parts of a cell.

1935    The concept of an ecosystem as a balanced interdependent network of relationships linking organisms and their environment is introduced by A. G. Tansley. The minimizing of competition by filling compatible "niches" is believed to be critical for the balance and functioning of ecosystems. Changes such as the introduction of a new species or an old species that develops a novel competitive advantage can upset the balance with potential catastrophic consequences. Opponents of the release of genetically modified organisms worry that precisely such a scenario will develop.

1938    X-ray scattering is first used to study the folded structure of DNA by physicists William Thomas Astbury and F. O. Bell. This technique, pioneered by English physicist Sir William Lawrence Bragg in 1912 for small molecules, also proves suitable to determine the structure, on the scale of the atom, of even large DNA and protein molecules.

        *Bacillus papilliae,* a bacterium, becomes the first microbial product registered by the U.S. Government as a control for Japanese beetles. Large-scale spraying over several states on the East Coast is the first such substantial release of microbes.

1939    French bacteriologist René Jules Dubos isolates the first antibiotic, gramicidin, from a common soil bacterium. Dubos later becomes a noted environmentalist.

1940    U.S. geneticist George Wells Beadle and U.S. biochemist Edward Lawrie Tatum, cowinners of the Nobel Prize in 1958, propose that one gene specifies one enzyme from studying the inheritance of traits in the common bread mold, *Neurospora crassa*.

1941    The term *antibiotic* is coined by Ukrainian-born U.S. microbiologist and Nobel Prize–winner Selman Abraham Waksman to describe compounds produced by microorganisms that kill bacteria to gain an advantage in the competition for nutrients and space. These toxins are designed to be very specific against the bacteria since the microorganisms producing them still have to live in the same environment. Pasteur had made the original observations of "antibiosis" in 1877.

        While publicly denouncing germ warfare, President Franklin Delano Roosevelt approves a secret plan to develop a U.S. biological warfare capability. By 1942 the United States has a four-pound anthrax bomb.

1944    DNA is recognized as hereditary material when microbiologists Oswald Avery, Colin MacLeod, and Maclyn McCarty transmit the virulence of a strain of *Pneumococcus* to a nonvirulent strain with isolated DNA alone.

1946    Gene mutation by chemicals with subsequent alteration of traits is observed by C. Auerbach and J. M. Robson. These observations provide the rationale for explaining genetic diseases.

        Transfer of DNA and genetic traits between *E. coli* strains by conjugation (bacterial mating) is demonstrated by two U.S. Nobel Prize winners, geneticist Joshua Lederberg and biochemist Edward Lawrie Tatum.

1950    Erwin Chargaff and coworkers demonstrate the pairing of purines and pyrimidines across the strands of the DNA helix: adenine (A) with thymine (T) [A=T] and cytosine (C) with guanine (G) [C=G]. This explains the constant ratios of these nucleic acids in DNA. This DNA sequence complementarity removes one of the last objections to nucleic acid encoding of hereditary information. These findings are instrumental in determining the physical structure of DNA three years hence.

1951    American geneticist Joshua Lederberg demonstrates that bacteria can exchange part of their genetic material. He calls the exchanged part a plasmid. He also finds that viruses that attack bacteria can act as an intermediate to transfer genetic matter between bacteria. Lederberg later becomes involved in recombinant DNA and biotechnology policy and debate in the 1970s.

1952    X-ray diffraction patterns of the ß form of DNA are described by English physicist Rosalind Franklin.

1953    Using U.S. biochemist Edwin Chargaff's findings on purine-pyrimidine pairs, the X-ray scattering data of Rosalind Franklin, and hand-built models, the double alpha-helical structure of DNA is described by English biochemist James Dewey Watson, physicist Francis Harry Compton Crick, and physicist Maurice Wilkins. Rosalind Franklin dies from cancer not long after the historical article is published.

Seymour Benzer, a U.S. physicist turned biologist, uses a technique called fine structure mapping to show that there are many sites within a single gene that are susceptible to mutation. This means that there are many ways a single gene can be altered and that the observed effect may not be the same for all mutations. It explains how genetic alleles can exist, that they are variants of the same gene, although each mutation maps slightly differently and sometimes shows a slightly altered trait.

1956    U.S. Nobel laureate biochemist Arthur Kornberg's discovery of DNA polymerase, the enzyme responsible for the replication of DNA, answers part of the question of how genetic information is faithfully copied for transmission to the next generation.

The Nobel Prize–winning work of American biochemist Christian Boehmer Anfinsen reveals that the three-dimensional structure of proteins is determined by the order of the amino acids in the protein chain. He shows that a string of chemically synthesized amino acids with the same sequence as the biological enzyme RNase A folds properly and expresses the catalytic activity of the native enzyme. Twenty years later molecular biologists take advantage of the natural folding of polypeptide chains when they begin to express nonbacterial proteins in large quantities in bacteria to study their function.

1957      The remarkable fidelity of DNA replication becomes evident when biochemists Matthew Meselson and Franklin Stahl show that replication proceeds by copying one of the original strands, using the A=T and C=G rules surmised by Erwin Chargaff. The replication proceeds in one direction only from one end (the 5' end) of the DNA and a proofreading activity of the DNA polymerase enzyme checks for mistakes.

1958      Francis Harry Compton Crick and Russian-born U.S. physicist George Gamow propose the "central dogma" of information flow in molecular genetics: DNA codes for RNA which codes for protein.

          Microbiologists Samuel Bernard Weiss, J. Hurwitz, and others report the discovery of DNA-directed RNA polymerase. This provides essential support for the hypothesis. Reverse transcriptase, an RNA-dependent DNA polymerase found in a family of retroviruses including some cancer viruses and the HIV virus linked to AIDS, proves an exception to this rule. The independent discovery in 1970 of this enzyme by cell biologists Howard Temin and David Baltimore proves of immense value to molecular biologists because it allows the cloning of proteins from their expressed messenger RNA (mRNA).

1959      French scientists Jerome Jean Louis Marie LeJeune, M. Gautier, and Raymond Alexandre Turpin discover an extra chromosome in the nuclei of cells from children with Down's syndrome. This unbalanced extra dose of genes results in mental retardation and a host of other malformations in this genetic disorder. It is a relatively common birth defect that occurs spontaneously with increasing frequency when maternal age is greater than 35 years.

          French Nobel Prize–winning biochemists François Jacob and Jacques Lucien Monod establish that the

1959     internal controls for gene regulation reside in the DNA
cont.    sequence on the chromosome as mappable features dis-
         tinct from the portion of genes coding for proteins. The
         steps in protein biosynthesis are also worked out at this
         time.

1961     The genetic code words for amino acids are identified
         over a period of five years in the labs of U.S. Nobel
         Prize–winning biochemists Robert William Holley,
         Marshall Warren Nirenberg, Har Gobind Khorana, and
         Severo Ochoa. As stated by the "central dogma" of mol-
         ecular biology the code is transcribed from DNA into
         an intermediate messenger RNA (mRNA), the function
         of which has been defined by Jacob and Monod. The
         mRNA is subsequently translated into protein using
         the genetic code words for the amino acids.

1963     The first textbook based on the principles of modern
         ecology is published by biologist E. P. Odin.

1964     The "Green Revolution" begins with the development
         of new strains of rice at the International Rice Institute
         in the Philippines. With sufficient fertilizer, yields of
         previous strains are doubled. While the increase is
         hailed as providing food for the hungry, issues such as
         increasing the dependence of the Third World on the
         developed industrial nations for fertilizers and the
         speeding of nutrient depletion of soils are downplayed.
         Acknowledgment of the impact of these issues comes
         later, with experience, and engenders significant debate
         over the introduction of genetically engineered crops in
         the 1990s.

         Genes coding for bacterial antibiotic resistance are
         found on plasmids, small circular pieces of DNA that
         remain separate from the main DNA of the bacterium.
         The most common antibiotic resistance genes code for
         enzymes that destroy the drug molecules before they
         can harm the bacteria. For example, the enzyme ß-lacta-
         mase hydrolyzes ß-lactam antibiotics such as penicillin.

1967     Polynucleotide ligase from *E. coli* is isolated and stud-
         ied by U.S. biochemists Samuel Bernard Weiss and

Charles Clifton Richardson. This enzyme is important in DNA damage repair, closing breaks in DNA by forming a new bond between pieces of the same strand of DNA. A bacteriophage T4 variant of the ligase is a routine tool in engineering recombinant DNA to splice together pieces of genes.

1970    The first restriction endonuclease enzyme, *E. coli* restriction endonuclease I (Eco RI), is isolated. Restriction enzymes, discovered in 1968, protect bacteria in their natural environment from foreign DNA. In the laboratory they make manipulation of DNA predictable and much easier by cutting at specific nucleotide sequences.

1971    General Electric and Indian-born U.S. microbiologist A.M. Chakrabarty apply, initially without success, to the U.S. Patent Office for a patent on oil-eating *Pseudomonas* bacteria created to deal with ocean oil spills. The patent is eventually issued after review by the Court of Customs and Patent Appeals. This landmark decision removes the distinction between patenting animate and inanimate inventions and opens the infant biotechnology industry to development. Finally, in 1980 the U.S. Supreme Court upholds the patent.

1972    The first recombinant DNA molecules constructed using restriction enzymes and DNA-ligase are fabricated at Stanford University. In 1973 restriction enzymes are used for cloning frog DNA into bacteria by the laboratories of U.S. biochemists Stanley Norman Cohen at Stanford and Herbert Wayne Boyer at the University of California, San Francisco. A patent issued to Stanford University and the University of California on this process collects a royalty for every cloning experiment performed with these enzymes.

1973    Public concern is expressed over the possible production of dangerous hybrid organisms by the new recombinant DNA technology. Arising from discussions at the Gordon Conference on Nucleic Acids, scientists draft a public letter that is published in *Science* on the possible biohazards of DNA splicing.

1973
cont.

Congress creates the Environmental Protection Agency (EPA) to act as the national watchdog for the environment.

1974

The National Institutes of Health (NIH) forms the Recombinant DNA Molecule Program Advisory Committee (RAC) on October 7. It is charged with framing guidelines for recombinant DNA research and reviewing gene therapy protocols. Its first meeting is in February, 1975. Critics of recombinant DNA technology call for a worldwide moratorium, which is respected, on certain kinds of experiments while the potential dangers are studied further.

1975

The first mammalian gene (the rabbit gene coding for production of globin, the protein part of the oxygen carrier hemoglobin) is cloned in bacteria by molecular biologists A. Efstratiadis, F. Kafatos, A. Maxam, and T. Maniatis.

The Asilomar Conference on Recombinant DNA Molecule Research is held to discuss concerns and progress in providing biological containment of experimental organisms and in assessing the safety of recombinant DNA research. Tensions run high as scientists feel that they are being railroaded by activists who don't understand or care about scientific issues or the freedom of inquiry. Nonscientists, on the other hand, feel that the issues at stake, such as the safety of the environment, are too important for scientists, who have their own agendas and livelihoods at stake, to be deciding on their own. The Senate Subcommittee on Health, Committee on Labor and Public Welfare begins the first public debate on recombinant DNA on April 22 with a series of hearings on genetic engineering chaired by Senator Edward Kennedy. Other nations seem to be following the lead of the United States in the controversy at this time. The Report of the Working Party on the Experimental Manipulation of the Genetic Composition of Microorganisms in the United Kingdom calls for special laboratory precautions for recombinant DNA research.

1976

The long-awaited publication of the NIH's first Federal

Safety Guidelines on Recombinant DNA Research on June 23 establishes a voluntary system of regulation for institutions receiving federal funding for research. The Department of Heath, Education, and Welfare–NIH guidelines are published (Federal Register 41(131): 27902–27943). Enforcement relies to a great extent upon existing statutes and regulatory powers. While restricting many categories of experiments, the guidelines do not go far enough for some people, while they are considered unnecessarily confining and bureaucratic to others. Industry is expected to comply voluntarily with the regulations and this disturbs many people who are concerned that the profit motive will override notions of safety in bringing products to market. The stringency of the guidelines applied to recombinant DNA research is later amended as experimental evidence accrues supporting the safety of some types of experiments. Commercial release of recombinant organisms into the environment is regulated by the network of government agencies empowered by the numerous statutes covering biological and chemical products.

Robert Swanson, an investment broker, and Herbert Boyer, a molecular biologist involved in the first genetic engineering experiments, found Genentech, the first company based on genetic engineering technology, to produce medically important molecules. Thousands of other such biotechnology companies will rise and fall through the years, but Genentech, the first and still the largest, markets some of the first genetically engineered biological products, mostly hormones or growth factors.

1977    A U.S. National Academy of Science meeting on recombinant DNA is held in Washington, D.C., on March 7–9. It is intended as a forum on industrial applications of the new genetic engineering technology. However, major figures such as Nobel Prize–winner George Wald and his wife Ruth Hubbard, both committed foes of recombinant DNA technology, take advantage of the media presence to support their cause. They speak to the audience, turning the panel discussions into a debate and encouraging dissenters in the audience, including the activist Jeremy Rifkin. Rifkin argues that

1977
cont.
the nature of life itself is at stake and that it is necessary for the people to keep "inhuman science" in its
place.

Sixteen bills are introduced in Congress to regulate recombinant DNA research; none are ever passed into
law.

The first recombinant DNA molecule incorporating
mammalian DNA is produced with genes for
(hemo)globin. The molecular basis for diseases with
special impact on ethnic populations—sickle cell anemia (African descent) and beta-thalassemia (Mediter-
ranean descent)—is shown by DNA sequencing
procedures developed the same year to be single amino
acid changes in these vital oxygen-carrying proteins.

1978
The Nobel Prize in medicine is awarded to Daniel
Nathans, Hamilton O. Smith, and Werner Arber for the
discovery and use of restriction enzymes for genetic engineering. A number of human gene products are produced by recombinant DNA techniques, including
somatostatin and insulin. The RAC is expanded to include members of the general public.

The first baby conceived by mixing human sperm and
egg outside of the body (*in vitro* fertilization, or IVF) is
born in the United Kingdom. Nineteen years later, a
sheep is cloned by transfer of an intact adult cell nucleus into an egg cell (ovum) whose nucleus has been
removed.

1979
The NIH guidelines for recombinant methods of handling viral DNA are relaxed with the accumulation of
safety data and the development of impaired host organisms and engineered vectors. Cancer-causing genes
are now studied by transformation of cultured cells
with DNA from malignant (cancerous) cells. Ironically,
these are the types of experiments that in the summer
of 1971 had set Robert Pollack and Paul Berg to contemplating the possible ramifications of recombinant
technology applied to cancer genes, leading to the
Asilomar conferences.

1980    The Nobel Prize in chemistry is awarded for creation of the first recombinant DNA molecules and for DNA sequencing methods. U.S. molecular biologist Kary Mullis and others at the Cetus Corporation in Berkeley, California, invent the polymerase chain reaction (PCR) technique of replicating selected DNA sequences from a mixture of DNAs. This methodology revolutionizes molecular biology in the 1980s. The patent for the PCR process is later sold to Hoffman-LaRoche, Inc., in 1991 for $300 million. Biogen, another of the early genetic engineering firms, is founded by U.S. molecular biologists Walter Gilbert and Charles Weissman. It begins producing interferon, a potent antiviral protein. Revised NIH guidelines are published, including voluntary compliance by non-NIH funded institutions (Federal Register 45[20]: 6724–6749, and revised Federal Register 45[227]:77384–77409).

1981    The first transgenic mammals are produced at Ohio University when foreign genes are transferred into mice. Golden carp produced by Chinese scientists are the first example of cloned fish. Sickle cell anemia becomes the first genetic illness to be diagnosed before birth at the gene level by restriction enzyme analysis of DNA. Congressman Al Gore begins a series of hearings on the relationship between academia and the growing commercialization of biomedical research. Such concerns are the harbinger of the university/industry web linking the many hundreds of new biotechnology companies with their academic founders and established industries. The fear that these interconnections will drastically influence the university research environment proves prophetic. The NIH publishes a final plan to formally assess the risks of recombinant DNA research (Federal Register 46[111]: 30772–30778).

1982    Human insulin produced by recombinant DNA methods is marketed by the pharmaceutical company Eli Lilly as Humulin®. In addition to relieving a forecast shortage of insulin for the increasing number of diabetics, allergic reactions to animal insulins are reduced by providing the human protein sequence. This hormone is the first of many biopharmaceuticals—small-to-

1982      medium-size proteins with potent biological effects on
cont.     particular cell types, previously available only in minute
          quantities—to be introduced into medical therapy.

          A foreign gene cloned into tobacco plants is success-
          fully transmitted to progeny in common Mendelian
          fashion, just like indigenous genes. Besides showing
          the similarity of genetically engineered changes to
          "normal" genetics, the transfer demonstrates the prac-
          ticality of engineering plant genes that soon become of
          major economic and ecological interest. The following
          year U.S. patents are granted to companies for geneti-
          cally engineered plants.

          A human bladder cancer gene cloned into E. coli is
          shown to be responsible for the cancer-causing activity
          of the protein when expressed in mammalian cells. It
          contains a single base pair change that results in a single
          amino acid change in the protein. This observation sug-
          gests that some cancers may be genetic diseases. AIDS is
          recognized as a syndrome, although the causative or-
          ganism is not identified until the following year.

          University of California scientist Stephen Lindow is the
          first to ask permission to deliberately release geneti-
          cally modified organisms into the environment to pre-
          vent frost damage on potatoes and strawberries,
          stimulating a storm of controversy. His protocol is ap-
          proved by NIH the following year.

1983      A chemical method for preparing synthetic genes is in-
          vented by chemist Marvin Carruthers. Leroy Hood's
          automation of the procedure at the California Institute
          of Technology makes gene synthesis a routine genetic
          engineering tool.

1984      Several moves are made to establish guidelines for con-
          trolling the impact of new medical technologies on hu-
          mans. Activist Jeremy Rifkin authors a resolution signed
          by fifty-six religious leaders and eight scientists sup-
          porting somatic cell modification as human gene ther-
          apy while opposing germ line treatments, which are
          transmissible to subsequent generations. The National

Organ Transplantation Act passes under the auspices of Senator Al Gore. It prohibits interstate commerce in organs or organ subparts for profit but allows transfers for medically relevant research purposes. The Warnock Report issued by the Parliamentary Committee of Inquiry in England recommends limiting research on human embryos.

1985     Congress authorizes the Biomedical Ethics Review Board to examine the ethical implications of genetic engineering of humans and to recommend policies and legislation to control potential abuses of the technology. The Biotechnology Science Coordinating Committee (BSCC) is established under Senator Al Gore as part of the U.S. Government Office of Science and Technology Policy.

Federal courts rule that private companies do not need NIH approval for field testing of genetically engineered organisms. EPA approval has, however, been required since 1984. MVP, a genetically engineered biopesticide developed by Mycogen, becomes the first such product approved by the EPA. It is a protein toxic to certain insects.

1986     Amid the controversy over relationships between universities and industry, CRADAs (Cooperative Research and Development Agreements) are instituted through the Technology Transfer Act of 1986 to facilitate technology transfer from government (NIH) to industry. Although this enables high-ranking government scientists to tap more private funds for research, the potential conflict of interest erodes public confidence in personal and government impartiality.

The first field trials of genetically engineered plants resistant to insects, viruses, bacteria are undertaken.

In a statement he has regretted since, Nobel Prize–winner Walter Gilbert calls the complete human genome sequence "the Holy Grail" of biology. Although his DNA sequencing chemistry is at the core of the determination of the complete human genome sequence, he

1986    later realizes that understanding of how that sequence
cont.   translates into biological function is the true goal.

1988    James D. Watson, then director of Cold Spring Harbor
        Laboratories where much of the new recombinant DNA
        technology has had its start, is appointed head of the
        National Center for Human Genome Research. It is felt
        that a scientist with the stature of a discoverer of the
        double helical structure of DNA is needed to galvanize
        the massive effort required to determine the complete
        DNA sequence of the human genetic complement, the
        genome. The first patent (U.S. patent 4,736,866) on a liv-
        ing transgenic animal, the Harvard OncoMouse®, is is-
        sued to Phil Leder and Harvard University. This mouse
        has been engineered to be highly susceptible to certain
        types of cancer; it is an animal model for antitumor ther-
        apies. The OncoMouse® patent is refused in Europe.

1990    While most critics worry about private industry com-
        mercializing genetic engineering, the first gene patents
        are actually obtained by the government's (NIH's)
        Craig Venter. He patents thousands of genes based on
        short pieces of DNA sequence whose functions can be
        predicted. Venter soon leaves NIH to start a biotechnol-
        ogy company outside of Washington, D.C., in Gaithers-
        burg, Maryland, based on these "expressed sequence"
        tags (ESTs).

        The Human Genome Project begins, a fifteen-year proj-
        ect to sequence the entire human genome.

        An Office of Technology Assessment reveals that 13%
        of Fortune 500 companies are using or have used some
        form of genetic screening in hiring decisions.

        Ashanthi DeSilva, a young girl with an adenosine
        deaminase (ADA) enzyme deficiency, receives the first
        cellular gene therapy using her own white blood cells
        genetically modified to regain their immune function.
        Modest improvement is seen, but most important, no
        ill effects are noted. Numerous other gene therapy pro-
        tocols soon follow, directed against intractable diseases
        such as cancer, Duchenne muscular dystrophy, and

cystic fibrosis as clinicians struggle to learn if the new technology will live up to its potential.

In the United Kingdom, the Human Fertilization and Embryology Act of 1990 bans human cloning.

1991    The Task Force on Genetics and Insurance is established at the National Center for Genome Research (NIH/Department of Energy) to address issues raised by genetic testing. This is part of the comprehensive plan to control consequences of the explosion of knowledge from the Genome Project.

1992    Wisconsin is the first state to forbid discrimination in employment or insurance based on an individual's genetic background. Other states start to follow suit. Loopholes remaining for company-financed insurance plans eventually lead to the introduction of protective Federal legislation in 1997.

1993    Despite heavy controversy, technology transfer continues unabated, with 200 CRADAs issued. Calgene markets genetically engineered delayed-ripening Flavr Savr/MacGregor tomatoes.

1994    Fifty-nine gene therapy protocols involving 150 patients are approved.

1995    The first complete nucleotide sequence of a bacterium (*Hemophilus influenzae*) is published. One hundred and six clinical protocols involving gene therapy have been approved and 597 patients have undergone experimental gene transfer since the first gene therapy experiment in September 1990. NIH is spending $200 million per year on gene therapy research. Biotechnology companies are thought to be spending as much again or more. The NIH Recombinant DNA Advisory Committee (RAC) concludes after a study of gene therapy that despite great promise and expectations, clinical efficacy has not been demonstrated definitely in any gene therapy protocol.

1996    The Recombinant DNA Advisory Committee (RAC) is disbanded and its functions taken over by various

1996    government agencies. The RAC has served through the
cont.   turbulent years of assessment of the safety of the new
        recombinant DNA technology while the regulatory as-
        pects were being developed.

1997    The cloning of a sheep from a somatic cell nucleus in
        Scotland raises a storm of concern over human cloning,
        previously ignored because it was so far from realiza-
        tion. Rhesus monkeys are reported to be cloned. Sud-
        denly human cloning appears imminent, although
        formidable technical obstacles remain. Federal legisla-
        tors rush to introduce a number of bills prohibiting
        human cloning. Protection against genetic discrimina-
        tion is also improved by the introduction of several ge-
        netic confidentiality and nondiscrimination bills.

        Although the public is willing to accept genetically en-
        gineered medical products, it is another matter entirely
        when it comes to what it eats. Reaction against geneti-
        cally engineered foods in the United States results in
        the Bovine Growth Hormone Milk Act regulating label-
        ing of milk products produced with the aid of the syn-
        thetic hormone. The European Union bans milk or meat
        products produced with synthetic growth hormone
        and maintains a similar stance on food products from
        genetically engineered plants. Flavr Savr tomatoes are
        removed from the market for "improvements."

1998    Human cloning remains a topical issue. The mouse, an
        important laboratory animal, is cloned by nuclear
        transplantation, a process that is expected to speed up
        certain kinds of research. The clinical and basic science
        community as well as supporters of medical research
        react to the pending blanket legislation against human
        cloning, calling for further study of the proposals. The
        intention is to avoid outlawing vital scientific and med-
        ical procedures needed to develop new therapies while
        maintaining a stance against cloning of human beings.

# Biographical Sketches 3

Although the roots of genetic engineering lie in the scientific tradition of the 19th century, the recombinant DNA revolution drastically accelerated in the late 1960s to early 1970s. The observations on heredity gradually acquired a biological basis in cells, nuclei, and chromosomes. Even if what genes were made of was unknown, their secret was suspected to be in the chromosomes. Chemical principles were shown to govern biological systems as well as the world around them. With this realization came the discovery of enzymes, the catalysts of chemical reactions inside cells. Soon the chemical substance DNA was determined to be the genetic material. The determination of the two-stranded double helical molecular structure of DNA and the deciphering of the code to translate DNA message into protein put the world on the verge of the revolution. It was the discovery of the utility of cellular enzymes to manipulate DNA that brought genetic engineering from a thought experiment into the real world.

This provided the possibility of medical, agricultural, ecological, and industrial breakthroughs, but carried a risk of potential harm to the environment. It led to a rethinking of issues of personal privacy and ethical precepts relating to how much control humankind should exert over nature and what it means to be human.

Progress in scientific understanding and in the application of the new technology has been largely stepwise, with many individuals taking part in the process. This chapter contains brief biographies of some of the men and women who contributed conspicuously to the development of genetic engineering. Biotechnology is a technical subject and recent; thus contemporary scientists are heavily represented, although politicians, an activist, and business people who figured prominently in the process are also included.

## William French Anderson (b. 1936)

WIlliam French Anderson grew up in Tulsa, Oklahoma, where he was born on December 31, 1936. He decided that he wanted to work on curing genetic diseases after finishing an undergraduate course at Harvard on DNA and genetics taught by James D. Watson, the Nobel laureate. After graduating with an A.B. in 1958, he studied with Francis Crick at Cambridge, receiving an M.A. in 1960. Then it was back to Harvard for an M.D. in 1963 followed by a pediatric residency at Boston Children's Hospital. His research career began in 1965 at NIH. There he studied hemoglobin synthesis and the inherited blood disorders thalassemia and sickle cell disease that result from mutant hemoglobin molecules. Anderson pioneered the use of iron binding drugs in iron overload conditions. In 1977 he became director of the Molecular Heredity Section at the NIH.

Searching for practical ways to introduce genes into cells, Anderson began to test retroviruses as potential therapeutic vectors, after removing their harmful genes. Retroviruses are a family of RNA viruses that include members that normally cause tumors and AIDS. The hemoglobin disorders proved too complex for beginning gene therapy in humans, so Anderson selected the newly discovered adenosine deaminase gene for replacement in patients immunocompromised by an ADA gene deficiency, a rare genetic disorder. Ater showing in 1988 that the engineered viral vector could survive in terminal human cancer patients, the first true gene therapy treatment was initiated by Drs. Anderson, R. Michael Blaese, and Kenneth W. Culver on September 14, 1990. The ADA gene in a retroviral vector was inserted into some of a nine-year-old girl's own white blood cells, which were then returned to her bloodstream.

Blood ADA levels rose and her immune system recovered partially—a success!

In 1991, Anderson founded the private company Genetic Therapy, to commercialize work that was being done in federal government laboratories and to provide vectors for gene therapy to hospitals and universities. He moved to the University of Southern California in 1992.

## Paul Berg (b. 1926)

Paul Berg is currently at Stanford University, where he has been on the medical center faculty since 1959. Born in New York City on June 30, 1926, he received his B.S. at Penn State in 1948 and his Ph.D. at Case Western Reserve University in 1952. Following postdoctoral studies at Washington University in St. Louis, he joined the faculty there in 1955. He remained there until he moved to Stanford as an associate professor in 1959. At that research mecca Berg found himself in the midst of scientific developments that would lead to recombinant DNA technology. He was recognized early on for his seminal contributions to the determination of the mechanism by which proteins are coded for by DNA. Before the age of 40, Berg was elected to the prestigious U.S. National Academy of Sciences, just one of a long series of awards that have acknowledged his contributions to science and society.

Berg's concern about the safety of certain kinds of DNA manipulation that he was contemplating, particularly tumor virus DNA, lead to the historic "Berg letter" to *Science* and the call for a moratorium on certain recombinant DNA research until safety issues were addressed. He was a key organizer of the Asilomar Conference in 1975 that led to the National Institutes of Health guidelines on recombinant DNA research issued in 1976. This was a historic instance of responsible self-regulation in scientific research. Paul Berg was awarded the Nobel Prize in chemistry in 1980 for his work on DNA. This recognition for his scientific accomplishments was followed by acknowledgment of his social contributions with the Scientific Freedom and Responsibility Award, the National Medal of Science, and the National Library of Medicine Medal.

His studies continued to involve recombinant DNA technology through the 1980s. In 1985 Berg was appointed to direct the Beckman Center for Molecular and Genetic Medicine at Stanford. Continuing his social responsibility ethic, in 1991 he was named as the head of the Human Genome Project Scientific Advisory

Committee of the National Institutes of Health. In this role he has been considering the ethical and practical issues raised by the new technology and the impact of genetic technologies on society.

## Herbert Wayne Boyer (b. 1936)

Herbert Boyer was born in Pittsburgh, Pennsylvania, on July 10, 1936, and attended tiny St. Vincent College in nearby Latrobe, Pennsylvania, where he received his A.B. in biology and chemistry in 1958. The former high school football lineman went on to the University of Pittsburgh, where he earned his M.S. in 1960 and finally his Ph.D. in bacteriology in 1963. After three years of postdoctoral experience at Yale University, he moved to the department of microbiology at the University of California at San Francisco in 1966. There he became embroiled in the early events of the recombinant DNA revolution. His laboratory was investigating the restriction enzymes of *Escherichia coli*, discovering that *E. coli* restriction endonuclease number I (Eco RI) cut DNA into reproducibly sized pieces, producing "sticky ends" that tended to adhere to other pieces of DNA treated with Eco RI. Boyer began collaborating with Dr. Stanley Cohen at nearby Stanford. Cohen had been studying plasmids, circular pieces of DNA carrying genes for such properties as antibiotic resistance in bacteria. The laboratories of Boyer and Cohen were soon able to recombine segments of DNA in a desired order to include genes for proteins, and to put the engineered plasmids back into bacteria. These altered bacteria, when properly grown, would then make that protein under direction of the plasmid DNA regardless of the origin of the DNA because the genetic code for the different amino acids was universal. This process was to provide the basis for the biotechnology industry.

In 1976 Boyer cofounded Genentech along with venture capitalist Robert Swanson, whose investment reflected his confidence in the new technology. He served as vice president of the company until 1990; he then shifted to the board of directors where he has remained to the present time. Genentech produced the first biopharmaceutical drug in 1978, the peptide hormone human insulin, which was subsequently licensed to the pharmaceutical company Eli Lilly. Genentech was the first biotechnology company to produce its own product when it launched human growth hormone in 1985. Boyer retained his appointment at the University of California at San Francisco, becoming a full professor in the genetics division of the department of Biochemistry and Biophysics in 1976. He has been the recipient of numerous academic and industrial honors and prizes for his pioneering applications of genetic engineering. Boyer is a member of the National Academy of Sciences, a fellow of the American Academy of Arts and Sciences, and a recipient of the Albert Lasker Basic Medical Research Award, considered by many to be the highest scientific award short of the Nobel Prize.

## Ananda Mohan Chakrabarty (b. 1938)

Ananda Chakrabarty was born in Sainthia, India, on April 4, 1938. He earned his B.Sc. from St. Xavier's College in 1958, and both his M.Sc. (1960) and his Ph.D. in biochemistry (1965) from Calcutta University in India. After a year as a scientific officer at Calcutta University, Chakrabarty moved to the United States and served as a research associate at the University of Illinois at Urbana until 1971. He then made the transition to industry, joining the General Electric Company in Schenectady, New York, as a staff microbiologist dealing with environmental pollutants. He recognized the potential for bioremediation of contaminants through the metabolic activity of microbes, and created strains of *Pseudomonas* bearing plasmids coding for enzyme activities capable of converting toxic oil spills to harmless byproducts. In 1971 General Electric and Chakrabarty filed a landmark patent case claiming specific genetically engineered organisms to clean up oil spills. The original application was rejected but was finally upheld by the Court of Customs and Patent Appeals in 1980. In 1979 Chakrabarty left General Electric to join the faculty in the department of microbiology at the University of Illinois Medical Center in Chicago, where he teaches and does research today. For his work with the oil-eating bacteria, Chakrabarty was recognized as Industrial Scientist of the Year in 1975 and later was honored with the Public Affairs Award from the American Chemical Society (1984) and the Pasteur Award (1991).

The U.S. Supreme Court turned back yet another challenge to the patent in 1990. This key patent decision removed the distinction between the patenting of animate and inanimate inventions. The possibility of protecting the commercial potential of genetically engineered organisms is a cornerstone of the biotechnology industry.

## John William Coleman (b. 1929 )

John Coleman was born in New York on December 30, 1929. As a black man in the 1940s, his educational choices were initially limited. He received his bachelor's degree from Howard University in 1950. He served as a physicist for the U.S. National Bureau of Standards from 1951 to 1953 and then as an instructor in physics at Howard University from 1957 to 1958. Coleman continued to advance his studies during this time, obtaining his master's degree from the University of Illinois in 1957. In 1958 he was hired as an engineer by the Radio Corporation of America (RCA), where he focused his research on the physics of electrons. During his tenure at RCA he earned his Ph.D. in biophysics at the University of Pennsylvania, receiving his degree in 1963. Although the electron microscope was invented in 1934 by Max Knoll and Ernst Roska, it remained a physics curiosity until 1940, when RCA demonstrated a crude commercial version. Coleman was involved in the development of the electron microscope at RCA as it was transformed from the behemoth 2.5-ton instrument of 1949, suitable for the study of materials science (metallurgy, ceramic surfaces), into a high-resolution device useable for properly treated biological samples.

## Francis Sellers Collins (b. 1950)

Francis Collins was born in Staunton, Virginia, on April 14, 1950. He was educated at the University of Virginia in Charlottesville, receiving a B.S. degree in 1970, an M.S. degree in 1972, and a Ph.D. in 1974 from Yale University. Collins earned his M.D. from the University of North Carolina at Chapel Hill in 1977 and then served his internship and residency at the hospital there through 1981. He continued his medical education as a fellow in medical genetics and pediatrics from 1981 to 1984 at the Yale University School of Medicine. He moved to the University of Michigan in Ann Arbor in 1984 as a faculty member in the departments of internal medicine and medical genetics, serving as chief of those departments from 1987 to 1991. He was also a Howard Hughes Investigator at the University of Michigan from 1987 to 1993. In 1993 he replaced the flamboyant and controversial James D. Watson as the director of the National Center for Human Genome Research at the National Institutes of Health (later upgraded to National Human Genome Research Institute) in Bethesda, Maryland.

Numerous honors and awards have been bestowed on Dr. Collins, and he serves on the editorial boards of several prestigious scientific journals dealing with human genetics, molecular biology, and genetic engineering. He is a member of the Institute of Medicine of the U.S. National Academy of Sciences.

While he was at the University of Michigan, Dr. Collins became known for cloning the most prevalent form of the defective cystic fibrosis (CF) gene, responsible for 80% of CF cases. This was seen at that time as a scientific coup and an example of the power of genetic techniques for elucidating the cause of diseases. It also raised the possibility of genetic therapy for this common inherited disease, found in as many as 1 in 2,000 Caucasian births. However, despite much effort, successful clinical treatment by gene therapy for this condition remains to be demonstrated.

Besides his work on CF, Dr. Collins is a key figure in the massive Human Genome Project, maintaining a constant presence in the public eye, balancing the swirling issues of ethics, morality, and privacy along with coordinating one of the most ambitious human endeavors ever, sequencing the human genome by the year 2006. Why would anyone take on this tremendous responsibility? Collins accepted a pay cut to direct the Genome Project, saying, "I feel I've been preparing for this job my whole life." Collins's energetic guidance, coupled with technical innovations, have moved up the official completion date to 2003.

## Francis Harry Compton Crick (b. 1916)

Francis Crick was born in Northhampton, England, on June 8, 1916. He trained in physics, receiving his B.Sc. from the University College of London in 1937. During and after World War II (from 1940 to 1947), he served the British Admiralty as a scientist. He was a medical research council student at the Strangeways Laboratory at Cambridge from 1947 to 1949 and was on staff at the molecular biology laboratory at Cambridge from 1949 to 1977. He began his famous collaboration with James D. Watson on the architecture of DNA, contributing his expertise in X-ray crystallography. This resulted in publication of the double helical model for DNA structure in 1953. Crick received his Ph.D. from Cambridge in 1954 and continued his work on the chemical basis for DNA structure. In 1955 he postulated that proteins were produced by adaptor molecules from the DNA code. By 1958 there was enough information to support the hypothesis that biological information flowed from DNA through RNA to protein structure, a hypothesis that became known as the central dogma of molecular biology. The eventual discovery of a group of viruses in which the genetic repository was RNA (hence the name retroviruses for some) required qualification of the hypothesis, but the concept still generally holds.

The 1962 Nobel Prize in physiology and medicine was awarded jointly to Crick, Watson, and Wilkins for their work in determining the structure of DNA and for the catalytic impact of this discovery on molecular biology and biochemistry. Many scientific honors followed over the years including election to the American, German, French, and Indian Academies of Science and becoming a fellow of the Royal Academy of Science.

In 1977 Crick moved to the Salk Institute for Biological Studies in La Jolla, California, and began to engage his interests in the neurosciences and complex behavior, as set forth in his book *The Astonishing Hypothesis* (Scribner, 1994). Crick's quiet style of doing science never earned him the antipathy that his more outspoken collaborator, James Watson, seemed to relish.

## Rosalind Franklin (1920–1958)

Rosalind Franklin was born in England on July 25, 1920. She graduated from the Cambridge University undergraduate program and began graduate work in physical chemistry. While still a graduate student, Franklin made her contribution to the World War II effort as an assistant research officer of the British Coal Utilization Research Association. From 1942 to 1946 she did pioneering work on coal microstructure and received her Ph.D. in 1945. In February, 1947, Franklin went to the Laboratoire Central des Services Chimiques de l'Etat in Paris, France, to learn X-ray diffraction techniques in order to further her microstructure work. After several productive years, she was eager to tackle biological structures. She took up John Randall's offer to set up an X-ray diffraction facility in Maurice Wilkins' laboratory at King's College, England, in 1951. She succeeded in obtaining high-quality diffraction patterns of oriented B-form DNA fibers, clearly recognizing the helical organization of the fiber and postulating a multichain helix with the phosphate backbone on the outside and the nucleic acid bases on the inside. This information was communicated at a departmental colloquium in November, 1951, at which James Watson was present. From this information, from Maurice Wilkins' communications, and from official King's College laboratory reports combined with their own data and intuition Watson and Crick built the models that culminated in the publication of the helical paradigm of DNA structure.

Although she did not receive full public credit for providing the experimental evidence of the structure of DNA, Franklin continued her work and published a structure for the A-form of DNA. In 1953 she left King's College for Birkbeck College in London to work with J. D. Bernel's group. There she produced one of the first X-ray analyses of the structure of the tobacco mosaic virus. Ironically, it was composed of protein subunits arranged in a hollow helix around a DNA core. Just a few years later, on April 16, 1958, Rosalind Franklin died of cancer.

Watson, Crick, and Wilkins shared the Nobel Prize for the structure of DNA in 1962. Many people attribute the apparent snub to the fact that Rosalind Franklin was a woman. It is true, however, that the Nobel Prize is awarded only to living persons.

## Albert Gore, Jr. (b. 1948)

Albert Gore, vice president of the United States in the Clinton administration, is a politician who has had a significant impact on the recombinant DNA debate through his involvement in ecological issues while he was in Congress. Born in Washington, D.C., on March 31, 1948, Gore obtained his A.B. degree from Harvard University in 1969 and then served a stint in Vietnam with the U.S. Army (1969–1971). Returning from the service, he did postgraduate work at the graduate school of religion at Vanderbilt University (1971–1972) and then at the law school there from 1974 to 1976. During this period (1971–1976) he was an investigative reporter and editorial writer for *The Tennessean*, where he honed his writing skills.

Gore entered public service in the U.S. House of Representatives (1977–1985), then served in the Senate (1985–1993). As a congressman he conducted a series of hearings on the connection between academia and industry in the commercialization of biomedical research. As a senator he was involved with the National Organ Transplantation Act regulating the commercialization of human organs for transplants. While in Congress he established himself as an expert on the environment and on nuclear arms control as well as bioethics. Gore is the author of *Earth in the Balance: Ecology and the Human Spirit* (1992).

# Leroy E. Hood (b. 1938)

Lee Hood was born in Missoula, Montana, on October 10, 1938. He received a B.S. degree from the California Institute of Technology (Caltech) in Pasadena, California, in 1960 and proceeded to the Johns Hopkins University School of Medicine in Baltimore, where he earned his M.D. degree in 1964. He then returned to Caltech, where he obtained his Ph.D. in biochemistry in 1968 and cultivated his interest in molecular immunology. From 1967 to 1970 he was a fellow at the National Cancer Institute studying immunology. Still drawn to California, Lee returned once more to Caltech, this time as a faculty member, where he rose to the rank of full professor by 1977. In 1989 he was appointed director of the National Science Foundation Center for Molecular Biotechnology. In 1992 he accepted the chairmanship of the new molecular biotechnology department at Caltech.

Hood's interest in the sequence basis for the polymorphic immune system made him impatient with the available technology for protein and nucleic acid sequence analysis. He began designing and building instrumentation that could automatically and rapidly determine the nucleic acid sequences of DNA and the amino acid sequences of proteins from very small amounts of sample. He also built machines that could chemically make synthetic nucleic acids (DNA and RNA) and polypeptides (proteins) from their component monomer units. He commercialized these instruments so that they became available to the scientific community and made rapid progress possible in sequencing the new genes being discovered daily. Without automated nucleic acid sequencing technology, the Human Genome Project would not be possible.

Dr. Hood is a member of the U.S. National Academy of Sciences. He was awarded the Louis Pasteur Award for Medical Innovation and the Albert Lasker Basic Medical Research in 1987 for his studies of immunological diversity. The technological advances spawned by the instrumentation and techniques developed in Hood's laboratory have earned him numerous other awards and honors. As genetic information accumulates from genome sequencing projects on multiple organisms, he is becoming increasingly involved in data analysis. He acknowledges that effectively utilizing this information to solve important medical problems is as big a challenge as acquiring the information in the first place.

# Edward Moore Kennedy (b. 1932)

Edward (Ted) Kennedy was born in Boston, Massachusetts, on February 22, 1932, to a prominent family steeped in politics and with a tradition of public service. The son of Joseph and Rose Kennedy, he was the brother of U.S. Attorney General Robert Kennedy and President John F. Kennedy, both victims of assassins' bullets. Kennedy obtained his A.B. from Harvard University in 1956 and pursued training in the law, with postgraduate study at the International Law School at The Hague, Netherlands, in 1958. He obtained his L.L.B. in 1959 from the University of Virginia. After passing the Massachusetts bar exam in 1959, he served as assistant district attorney for Suffolk County from 1961 to 1962.

Kennedy was elected U.S. senator for Massachusetts in 1962, and he remains an outspoken liberal voice in that body on social issues. During his tenure in the Senate he has served on the Judiciary Committee (1979–1981), the Armed Services Committee, and the Democratic Steering and Organization Committee. He has been particularly involved with health issues and was a prominent figure in Congress as chairman (1971–1980) of the Subcommittee on Health of the Labor and Human Resources Committee during the early attempts to legislate regulation of the use of recombinant DNA technology. He is also a member of the Biomedical Ethics Board. His publications include *Decisions for a Decade* (1968), *In Critical Condition: The Crisis in America's Healthcare* (1972), *Our Day and Generation* (1979, with Mark O. Hatfield), and *Freeze: How You Can Help Prevent Nuclear War* (1979).

# Michael Martin (b. 1956)

Michael Martin was born in 1956 in the rural Alabama town of Orrville and grew up on a small cotton and livestock farm. Martin's encounter with the southern corn blight of 1970, which destroyed his hopes of winning a Future Farmers of America award, left him searching for an explanation of why some corn varieties survived and others were wiped out. A county extension agent explained the basis of genetic resistance to the black teenager, setting him onto a career in crop improvement. Martin earned a B.S. in agronomy from Alabama A&M University and went on to Iowa State for an M.S. and eventually his Ph.D. (1982) in plant breeding and cytogenetics.

One of his former professors hired the newly graduated

Martin as a research center manager for the Garst Seed Company, a leader in hybrid corn development. Garst merged with Zeneca Seed of Britain in 1990 and continued to grow. Martin led Garst into the first commercial introduction of a herbicide-resistant as well as multigenic disease-, insect-, and climate-resistant corn strains. By 1996 he was supervising a staff of 247 with an annual budget of $18.5 million. The formation of Advanta Seeds by Zeneca and the Dutch Royal Vander Have Group further expanded his horizons and responsibilities. As the world's largest provider of canola and sunflower seeds, and ranking in the top five worldwide in corn seeds, Advanta Seeds had projected annual sales in 1997 of over $500 million. Martin is responsible for developing and implementing a research, sales, and distribution plan in his specialty of dent field corn and tropical corn varieties.

## Barbara McClintock (1902–1992)

Barbara McClintock was born June 16, 1902, in Hartford, Connecticut. After graduating from high school at the age of 16, she worked at an employment agency until she entered Cornell University's Agriculture College. She decided to study genetics and received her B.S. degree in 1923. She was fascinated by the chromosomes of corn (maize) plants; maize was the genetic system studied at Cornell at the time rather than the fruit fly, *Drosophila melanogaster,* which was in vogue at other universities. Careful microscopic study of maize development allowed McClintock to clearly see the chromosomes divide and rearrange. The characterization of this system became her Ph.D. thesis project, which she earned through the Department of Botany in 1927. She remained as an instructor until 1931 when she and Harriet Creighton published work that connected the physical interchange of maize chromosome material with the exchange of genetic information, just weeks ahead of the publication of similar observations with *Drosophila* by other groups. Very much a loner and uncomfortable with the norms of scientific behavior, McClintock held various positions and did research at Cornell, the University of Missouri, and Caltech between 1931 and 1940. Her reputation as a cytogeneticist was impeccable. In 1941 she was invited to join the staff of the Carnegie Institute of Washington (D.C.) Department of Genetics at the Cold Spring Harbor Laboratories on Long Island, New York, where she remained for the rest of her life.

McClintock was elected to the U.S. National Academy of Sciences in 1944, the third woman so honored. The subsequent year she was elected president of the prestigious Genetics Society of America, the same year that genes were demonstrated to be made of DNA. In 1951 McClintock tried to explain her observations on the transmission of certain maize characteristics by postulating that DNA could be rearranged on the scale of genes, but she failed to gain the general acceptance of the scientific community. A number of years later after molecular techniques demonstrated similar events with bacteria and viruses, McClintock's data on evolutionarily more advanced organisms obtained with nineteenth century equipment—a regular microscope, cross-breeding experiments, and careful observation—finally gained the recognition they deserved. DNA can rearrange; "jumping genes" are a biological reality. Similar DNA rearrangements within immune system genes in white blood cells are crucial for the construction of antibodies that protect against foreign invaders.

McClintock was honored for her contributions, receiving several awards, including the National Medal of Science in 1970, and finally winning the Nobel Prize in physiology and medicine in 1983. With all of the attention so richly deserved, Barbara McClintock remained quietly aloof, working in her laboratory and cornfields until she died on September 2, 1992.

## Louis Pasteur (1822–1895)

Louis Pasteur was born on December 27, 1822, at Dole in the Jura region of France. An unremarkable student until near the end of his secondary school career, Pasteur obtained his bachelor's degree in 1840, after a series of fits and starts, from the Collège Royal de Besançon. He eventually entered the École Normale Supérieure in Paris, receiving his doctorate in physics and chemistry in 1847. He continued to work at the École Normale while awaiting a faculty position and became involved in studies on the optical activity of molecules. This work became the starting point for his career-long involvement in diverse and seemingly unrelated research topics. In 1849 he became professor of chemistry at Strasbourg and then, in 1854, professor and dean of the faculty of science at Lille in the industrial region of France.

Associated with an institution closely allied with manufacturing interests, Pasteur found an outlet for his highly practical inclinations in Lille. His investigations into the biological and chemical origins of optical activity and his development of techniques for separating the components led Pasteur into the examination of fermentation reactions of various types. He conducted a series of experiments that in 1860 eliminated spontaneous generation as a possible explanation of spoilage plaguing the wine and vinegar industries. Pasteurization, the heat treatment method that bears his name, proved effective for several industrial applications, and he eventually patented the procedure in 1865. In 1857 he moved back to the École Normale as director of scientific studies. He was asked by the French silkworm industry to find a solution to a mysterious disease that was ruining silk production. Through long and tedious experimentation, Pasteur was able to demonstrate two simultaneously occurring microbial diseases and to work out procedures to eliminate further outbreaks.

During this time the germ theory of disease was receiving experimental support from Robert Koch, the father of bacteriology, and from others who, along with Pasteur, developed methods of growing pure cultures of disease-causing organisms. Joseph Lister introduced antiseptic surgery in which instruments, hands, and surroundings were chemically sterilized with carbolic acid in the 1860s. These precautions were immediately effective in reducing hospital patient mortality.

Pasteur began studies on anthrax, a deadly bacterial disease. He had to hire assistants to perform the necessary biological experiments since he, as an antivivisectionist and partially paralyzed by a stroke in 1868, could not do them himself. He became one of the pioneers of immunity and prophylaxis—that is, of preventing disease by immunization—following up on Edward Jenner's successful smallpox vaccinations in the late eighteenth century. A successful, highly publicized field trial of an attenuated anthrax treatment on susceptible farm animals in 1881 led Pasteur to extend his work to fowl cholera, hog cholera, and rabies. The July 6, 1885, treatment of nine-year-old Joseph Meister, who had been bitten by a rabid dog, with spinal cord extracts infected with attenuated virus prevented the inevitably fatal progression of the rabies disease and solidified Pasteur's fame.

In 1888 Pasteur became director of a new institute bearing his name, which was dedicated to the treatment of rabies and the development of more effective vaccines. Increasingly incapacitated by illness, Pasteur continued directing research at the institute. He died near Paris on September 28, 1895, and was honored with a state funeral at the Cathedral of Notre Dame. Pasteur was the recipient of numerous awards and much honorary recognition, including election to the Royal Society, to the Académie de Medécine, and to the Académie Française. His contributions both to the experimental demonstration of the germ theory of disease and to the practical development of treatments for microbial diseases laid the basis for modern immunoprophylaxis and therapy.

# Jeremy Rifkin (b. 1945)

Growing up in the late 1940s and early 1950s on Chicago's South Side, Jeremy Rifkin gave little hint of the political and social activist he would become. The son of a plastic-bag manufacturer and a charity worker, he developed his antiwar and civil rights sentiments at the University of Pennsylvania's Wharton Business School of Finance. He feels that growing up in the Vietnam War protest era steered him to his activist profession. Rifkin served as an organizer of the 1968 March on the Pentagon, and in 1969 founded the Citizen's Commission to focus on alleged U.S. war crimes in Vietnam. In 1971 he founded the People's Bicentennial Commission as a countercultural alternative to official government plans for the U.S. bicentennial celebration.

By the time the recombinant DNA controversy came on the scene, Rifkin was well entrenched as a foe of the establishment. He founded the nonprofit Foundation on Economic Trends, based in Washington, D.C., to influence government policy on a spectrum of economic, environmental, scientific, and technological issues. Although Rifkin lacked formal training in the sciences, beginning in the late 1970s he became a strident voice in opposing genetically engineered crops, patenting of genes, genetic engineering of animals and animal breeding practices, biological weapons, and the sale of foods modified by recombinant DNA. Almost exactly 20 years before the 1997 announcement of the cloning of the sheep Dolly by nuclear transfer, Rifkin and a group of protesters disrupted a National Academy of Sciences meeting on applications of recombinant DNA technology chanting, "We will not be cloned!" Through his foundation he has filed lawsuits against various government agencies, including the Department of Agriculture, on a number of fronts. He has challenged the use of genetic engineering in animal breeding on the basis of animal rights. On the grounds of inadequate environmental impact assessment, he has disputed various proposed releases of genetically modified organisms such as engineered frost-preventing bacteria (Frostban®) and crude oil-eating bacteria for oil spills.

Rifkin moved on to other concerns, including global economics, workers' rights issues, the impact of information technologies on the workplace, and global warming. However, in the 1990s his foundation began rallying grassroots opposition to the use of genetically engineered bovine growth hormone to enhance milk and meat production in cattle. In May, 1995, he orchestrated the issuance of a statement from 180 religious leaders from some 80 different religious groups, including United Methodist, Southern Baptist, Jewish, and Muslim organizations. It called for a moratorium on patenting of genetically engineered animals and of human genes, cells, tissues, organs, and embryos. In April 1997 the foundation, in concert with others, organized a global protest to oppose genetically engineered foods, cloning, and patenting of genes and life forms. Despite his fervent opposition to many aspects of genetic engineering, Rifkin maintains that he has never opposed biotechnology for genetic screening, for applying genetic knowledge to preventive medicine, or for the production of pharmaceuticals.

By enlisting strong emotional opposition to genetic engineering and other issues, Rifkin has incurred the enmity of government officials, scientists, and industry executives, to whom he is known as the "abominable no man." He responds by saying that critical perspective is needed in commercialization of genetic technologies, pointing to the lack of such a public debate on the nuclear and chemical technologies that led to the Three Mile Island, Chernobyl, and Bhopal disasters.

Rifkin is president of his Foundation on Economic Trends, president of the Greenhouse Crisis Foundation, and head of the Beyond Beef Coalition. He is the author of 13 books, including *Algeny* (Viking Press, 1983). This expresses a naturalistic or vitalistic view of nature; it caused a stir by questioning the objectivity and validity of Darwinian evolution using creationist arguments. Rifkin returned at the close of the 1990s to stir the pot on genetic engineering as the author of *The Biotech Century: Harnessing the Gene and Remaking the World* (Putnam, 1998).

## Marian Lucy Rivas (b. 1943)

Marian Rivas was born on May 6, 1943, in New York City. After receiving her B.S. degree from Marian College in 1964, she continued her studies at Indiana University where she earned an M.S. (1967) and a Ph.D. (1969) in the then new field of medical genetics. She extended her training with a fellowship at Johns Hopkins University from 1969 to 1971. Her first faculty position came at the Douglas College of Rutgers University in New Jersey, where she was an assistant professor from 1971 to 1975, just as the recombinant DNA revolution was beginning to sweep biology.

The medical applications and consequences of the new technology were to become apparent just a few years later. In 1975 Rivas moved to the hemophilia center of the Oregon Health Sciences University, where she has been a full professor since 1982. In 1978 she also became an associate scientist at the Neurologic Science Institute of the Good Samaritan Hospital. As a medical geneticist, Rivas has served on several committees on genetics at the National Institutes of Health. Her career has spanned human gene mapping and investigation of the genetic aspects of epilepsy. She has been involved in genetic counseling of patients and with the application of computers in clinical genetics.

## Maxine Frank Singer (b. 1931)

Maxine Singer was born in New York City on February 15, 1931. She received her A.B. from Swarthmore College in 1952 and her Ph.D. from Yale University in 1957. Singer began her career at the NIH as a Public Health Service Fellow and rose through the ranks, working on nucleic acid chemistry and metabolism through 1987. Her work on the biochemistry of animal viruses in the 1970s brought her into contact with the issues that she would address as cochair of the Gordon Conference on Nucleic Acids in 1973. She was instrumental in drafting a letter, signed by most of the participants at that conference, to the National Academy of Sciences and National Institutes of Medicine expressing concern about the safety of the new recombinant DNA technology. This public anxiety of a group of scientists combined with similar sentiments from Paul Berg eventually led to a moratorium on certain research with recombinant DNA proposed at the Asilomar Conference of 1975. She continued her involvement in both recombinant DNA research and in monitoring scientific social responsibility on the safety of recombinant DNA systems. In the 1980s, when the hysteria over the new technology had died down without ecological disaster, Singer began emphasizing the benefits to be derived from recombinant DNA applications. In the 1990s she pointed out that the Flavr Savr tomato genetically engineered to delay softening upon ripening was little different from tomatoes derived by years of cross-breeding and was just as safe to eat.

Singer's accomplishments were recognized by numerous awards including her election to the U.S. National Academy of Sciences and National Institutes of Medicine. She has been an emeritus scientist of the NIH National Cancer Institute and the president of the Carnegie Institute in Washington, D.C., since

1988. She supervises the First Light Program designed to interest girls and boys in science.

## Robert Swanson (b. 1947)

Robert Swanson made his impact on genetic engineering by providing the primary ingredient that was to allow this technology to become a major force today: investment capital. With an undergraduate degree in chemistry from the Massachusetts Institute of Technology (MIT) and an M.S. from MIT's Sloan School of Business, and fortified with four years of experience as an investment banker with Citibank, he was aptly suited for bridging the promise of the science with the principles of business competition and return on investment on which many a biotechnology startup company has foundered.

In 1975 Swanson found Herbert Boyer at the University of California, San Francisco. Boyer's and Stanley Cohen's labs were using the new restriction enzymes to cut the DNA of bacterial plasmids, leaving sticky ends. They used these sticky ends to recombine genes on plasmids, inserting them into bacterial cells for the production of specific proteins. Assured by Boyer that commercial application of genetic engineering was feasible, in 1976 Boyer and Swanson convinced Thomas Perkins of Kleiner-Perkins, a Silicon Valley venture capital firm, to advance seed money to establish Genentech. By 1978 Genentech had produced genetically engineered human insulin in bacteria and licensed the technology to Eli Lilly, a major pharmaceutical company and producer of animal (porcine) insulin for treatment of diabetes. In 1985 Genentech became the first startup biotechnology company to introduce its own biopharmaceutical product, human growth hormone. It remains a large, successful biotechnology company both on the business side and in scientific enterprise. Swanson was director and chief executive officer (CEO) from 1976 until 1990, when he was named chairman of the board.

## J. Craig Venter (b. 1946)

Craig Venter was born on October 14, 1946, in Salt Lake City, Utah. A year spent in Vietnam as a medical corpsman in 1967 helped galvanize his career plans along medical lines. In 1972 he obtained his B.A. degree from the University of California, San Diego, where three years later, in 1975, he earned his Ph.D. in physiology and pharmacology. After a research fellowship from 1975 to 1976 in cardiovascular pharmacology at the University of California, San Diego, he moved to the State University of New York at Buffalo, New York, where he was a faculty member in pharmacology and biochemistry until 1984. He then joined the National Institute of Neurological Disorders and Stroke at the National Institutes of Health in Bethesda, Maryland.

From his initial studies on pharmacologically important hormone receptors, Venter entered the then-infant field of genomics. He promoted a sequencing strategy called the expressed sequence tag/complementary DNA (EST-cDNA) technique to identify the active segments of human genes while skipping the inactive (untranscribed) parts of the genomic DNA sequence. The NIH raised a storm of scientific controversy when it sought to patent the EST gene fragment sequences without knowing the function of the proteins coded by the sequences. Many critics believed that patenting gene sequences would discourage research on the gene products. This criticism added to the controversy over whether genes should be patentable at all. With the growing ability to predict the function of new sequences based on characteristic structural and functional motifs of protein sequences, the EST-cDNA technique has been useful in chromosomal mapping projects. This is because the ESTs are large enough (about 300–400 base pairs) to represent unique sequences that can be rank-ordered along the length of the chromosomes.

In 1992, after failing to obtain expanded research funding for his EST sequencing project, Venter once again precipitated controversy by jumping from NIH to become the president and director of a private, nonprofit research center, The Institute for Genomic Research (TIGR), in Gaithersburg, Maryland. TIGR was allied with a biotechnology affiliate Human Genome Sciences (HGS), which was to commercialize selected sequences discovered by TIGR. Access to a cDNA database jointly maintained by TIGR and HGS and preferentially available to pharmaceutical companies supporting their research has been a bone of contention. Much of the disputed data—some 45,000 DNA sequences—has recently been made publicly available over the World Wide Web. Such debate over public and private interests has been a common feature of biotechnology research. It has been at once an ethical challenge for modern society and the driving force for the application of the new knowledge for the benefit of humankind.

# James Dewey Watson (b. 1928)

James Watson was born on April 6, 1928, in Chicago. He entered the University of Chicago at the age of 15 and received his B.S. degree in 1947. His applications for graduate study were rejected by Harvard and California Institute of Technology—ironically two places where he would later serve on the faculty—so he went to Indiana University, where he earned his Ph.D. in genetics in 1950. After postdoctoral study in Copenhagen (1950–1951), Watson joined the Cavendish Laboratory in Cambridge, England, where he met Francis Crick and became involved in the search for the structure of DNA.

As recounted in the autobiographical *The Double Helix* (Atheneum, 1968), he and Crick fashioned a model of two DNA strands wound around each other in a helical configuration held together by the pairing of DNA bases. Data for the model came from the X-ray crystallography and electron microscopic studies of Rosalind Franklin and Maurice Wilkins as well as from their own work. The DNA double helix model was published in 1953 and served as a cornerstone for nucleic acid biology in the infant fields of molecular genetics and biochemistry. For the far-reaching consequences of this paradigm, Watson, Crick, and Wilkins were jointly awarded the Nobel Prize in physiology and medicine in 1962.

Many scientific awards followed Watson's seminal work, including election to the U.S. National Academy of Sciences. Watson was a faculty member at Cal Tech from 1953 to 1955 and at Harvard from 1956 to 1976. He continued to contribute to the understanding of the triplet code of DNA by which the amino acid sequences of proteins are specified. In 1968 he became director of the Cold Spring Harbor Laboratory, where he continues to administer and advocate for basic research in science.

When the initiators of the Human Genome Project needed a prominent, forceful champion for their huge, collaborative DNA sequencing project, they turned to Watson, who provided the required spark and leadership from 1988 to 1990 to get the project rolling, pointing out its potential medical applications. He once commented, "We used to think that our future was in the stars. Now we know that it is in our genes." Always outspoken and, many felt, often abrasive, James Watson proved to be an able administrator and a positive force for the application of gene science. A staunch advocate of unfettered intellectual pursuit and the scientific method, he has been criticized for not paying enough attention to the impact of the Genome Project on political, social, or ethical issues.

# Facts, Data, and Opinion 4

## Statistical Data

In Western society, the recombinant DNA revolution has been subject to the same resistance to new technology that has commonly been seen throughout history. Analysts cite a variety of theories for this behavior, with fear of the unknown being a common motivation. A survey was conducted over the period from 1983 to 1995 to assess changes in the attitudes of the American public toward organized science. It indicated that as people become more accustomed to technological advances, they become less concerned with the rate of change and tend to feel that those changes contribute positively to their lives (Table 4.1).

Influential factors in the recombinant DNA debate include the potential impact on world ecosystems and on health, and personal control of information. Many other major controversies with implications for public health—such as locating the Seabrook nuclear power plant, drilling for offshore oil, landing of the Supersonic Transport aircraft, regulating the sale of the artificial sweetener saccharin, establishing standards for genetic carcinogens, and reducing ozone destruction by fluorohydrocarbons such as Freon®—

## Attitude toward
## Organized Science

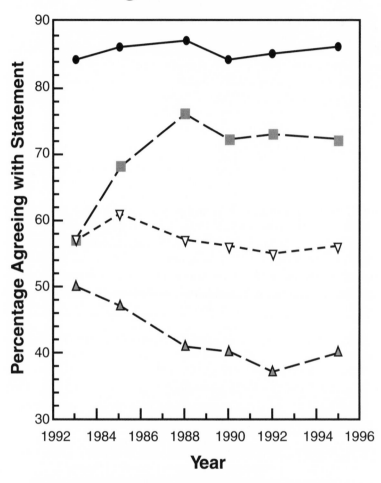

*Source:* Revised from National Science Board. *Science and Engineering Indicators—1996.* Washington, DC: U.S. Government Printing Office, Appendix, Table 7–20.

have been decided in the public forums of the courts, by legislation, and in public hearings of government agencies. Scientific input was only part of the process.

The genetic engineering controversy, particularly during the initial stages of the debate, has been distinct on several levels. Recombinant DNA technology is largely a scientific research tool, and scientists, a decided minority of the population, initially would have the most to gain from its application. The potential hazards, though widely discussed, had not been confirmed to occur, nor was the magnitude of the expected effects known. Further, in addition to the environmental and health issues, the public debate included ethics. Even if personal rights of privacy of genetic information and fair treatment are guaranteed, is it proper to manipulate the genetic substance in the first place? Achieving consensus on such issues of faith is a daunting prospect. A survey in the United Kingdom (Lee, Cody, and Plastow, 1985) found that 70% of the respondents found genetic engineering to be "morally wrong"; 62% found it "unnatural," and 27% found it "frightening." Similar responses were obtained in the U.S. (Hoban and Kendall, 1992). The beliefs about the ethical scruples also depend on the proposed use of the genetic engineering. In general, people are far more likely to consider genetic options permissible for medical conditions than for food enhancements or increased industrial production.

Outside of the gut feelings of moral rightness and trepidation toward change, two major contributors to the general public unease with genetic engineering are the highly technical nature of the arguments used to justify the safety of the technology and the magnitude of the harm that could be caused. This is played out through the general scientific illiteracy of the American public, particularly their lack of understanding of the process of scientific inquiry—how scientists arrive at conclusions, and what those conclusions mean. A 1995 survey by the Office of Technology Assessment concluded that 64% of adults surveyed did not understand the process by which measurements are made and comparisons drawn in experimental studies to determine which of two alternative treatments is better than the other and whether the difference is significant. Formal education, particularly science and math education, improved understanding of the process but still, over one-third (36%) of respondents with a high level of science/math education did not understand basic scientific proof (Table 4.2).

### TABLE 4.2
#### Public Understanding of the Nature of Scientific Inquiry (1995)

Respondents were presented with the following situation: "Two scientists want to know if a certain drug is effective against high blood pressure. The first scientist wants to give the drug to 1,000 people with high blood pressure and see how many experience lower blood pressure levels. The second scientist wants to give the drug to 500 people with high blood pressure and not give the drug to another 500 people with high blood pressure and see how many in both groups experience lower blood pressure. Which is the better way to test this drug? Why is it better to test the drug this way?"

| Level of understanding | A | B | C | D | Sample size |
|---|---|---|---|---|---|
| | | (% of sample) | | | |
| All adults | 2 | 21 | 13 | 64 | 2,006 |
| | | | | | |
| Formal Education | | | | | |
| Less than high school | 0 | 4 | 7 | 89 | 418 |
| High school graduate | 1 | 18 | 15 | 66 | 1,196 |
| Baccalaureate | 6 | 44 | 13 | 37 | 260 |
| Graduate/professional | 10 | 49 | 12 | 29 | 132 |
| | | | | | |
| Science/mathematics education | | | | | |
| Low | 0 | 9 | 12 | 79 | 1,125 |
| Middle | 3 | 30 | 16 | 51 | 530 |
| High | 7 | 45 | 12 | 36 | 352 |
| | | | | | |
| Sex | | | | | |
| Female | 2 | 20 | 13 | 65 | 1,053 |
| Male | 2 | 22 | 12 | 64 | 953 |
| | | | | | |
| Attentiveness to science or technology | | | | | |
| High | 5 | 34 | 14 | 47 | 195 |
| Medium | 3 | 22 | 13 | 62 | 946 |
| Low | 1 | 16 | 13 | 70 | 865 |

A = Understands science as the development and testing of theory.
B = Not A level understanding, but understands concept of experimental study, including meaning and use of a control group.
C = Not B level understanding, but understands science to be based on careful and rigorous comparison.
D = Does not understand science on any of the above levels.

*Source:* Revised from National Science Board. *Science and Engineering Indicators—1996,* Table 7–9, p.304.

The survey showed little correlation between people's educational level (from non-high school graduate to baccalaureate) and their feelings about whether the benefits to be derived from genetic engineering were greater than the risks. If anything, non-high school graduates were slightly more optimistic about the benefits than those with higher education (Table 4.3).

Early fears about genetic engineering focused on the probability of the escape of modified organisms from laboratories, causing

Table 4.3
# Public Assessments of Genetic Engineering
### Educational Status

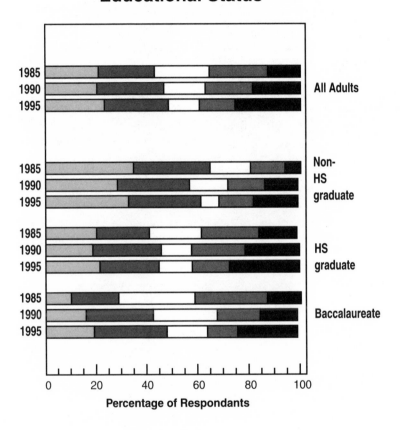

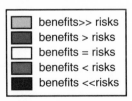

All Adults

Non-
HS
graduate

HS
graduate

Baccalaureate

Percentage of Respondants

- benefits>> risks
- benefits > risks
- benefits = risks
- benefits < risks
- benefits <<risks

*Source:* Revised from National Science Board. *Science and Engineering Indicators—1996.* Washington, DC: U.S. Government Printing Office, Appendix, Table 7–23.

human disease and disruption of the environment. Initially, little experimental evidence was available to drive bona fide risk analysis; even the scientists were having to make quite unsophisticated estimates. The fantastic scenarios of massive plagues faded from discussion as data became available and as biocontainment improved, through the development of standard workhorse organisms with drastically reduced ability to survive outside of the laboratory. There remain serious questions about the safety of the large-scale deliberate release of organisms into the environment for agricultural or bioremediation applications. These questions are presently addressed on a case-by-case basis through the EPA and allied agencies.

Surveys by the Office of Technology Assessment published in 1987 indicate that the public opinion is less than mollified. Respondents made little distinction between specific dangers that might be expected from the use of genetically altered organisms in the environment (Office of Technology Assessment, 1987). They were asked about approving agricultural use of an organism with no risk to humans but with increasingly remote possibilities, ranging from high to unknown to one in a million, of losing some local species of plant or fish. A substantial percentage of respondents (18%) would not approve even at the lowest risk (1:1,000,000). Significantly, having available some quantitative measure of risk, such as 1:1,000 or 1:10,000, provided a basis for a decision. Almost half (46%) of respondents would not approve the use of the organism if the risk were unknown, even if it were "very remote." The specific proposed use of the organisms affected the likelihood of public approval; release of disease- and frost-resistant crops or "oil-eater" bacteria were more likely to be approved than more effective biopesticides or larger engineered game fish.

While people said they wanted to be consulted on whether or not genetically engineered organisms should be released into the environment, when it came to the actual decisions on whether commercial firms should be allowed to release genetically modified organisms on a large scale, they were more comfortable delegating the decision to a government agency or else to an external scientific body. The distribution of opinion was similar across different political affiliations. Interestingly, only 5% or less of respondents believed that the public at large or communities should decide. Surveys asking most of the same questions over 10 years later have been conducted in other countries including New Zealand and Japan (Macer, 1998), the European Union (Biotechnology and the European Public Concerted Action Group, 1997), and Canada (Eisiedal, 1997). Despite the differing extent of involvement in biotechnology, different cultural traditions, and the intervening years since the U.S. survey, the opinions of the sampled populations were remarkably consistent in their basic mistrust of biotechnology, fuzzy public understanding of questions of relative risk and benefit, apprehension about effects on the environment, and the desire to be consulted in deciding what kind of work should be allowed. There apparently remained a general lack of confidence in the effectiveness of government regulation and in the motivations and accountability of industry. People were far more willing to see genetic engineering used in health-related ways than for food production, and all viewed the potential for human cloning research negatively.

# Genetic Engineering as a Business

Genetic engineering has been a vastly enabling technology, both at the level of being able to do things that were previously impossible or completely novel and in expediting and improving the capabilities of current practices. Many believe that it has been so successful and developed so rapidly, especially in the United States, primarily because obstacles to its commercialization were removed at an early stage. The General Electric Company/ Chakrabarty patent decision from the Court of Customs and Patent Appeals during the early 1970s was a landmark determination. It allowed the patenting, as a new form of life, of bacteria engineered to metabolize oil spills. Previously, certain types of plants were the only form of life that could be patented. Eventually upheld by the Supreme Court in 1980, the Chakrabarty patent plus an additional patent to Stanford University (Drs. Stanley Cohen and Herbert Boyer) on gene splicing methodology, issued the same year, laid the legal groundwork for the biotechnology industry. In 1987 the first patent was granted to Harvard University on a genetically engineered mammal, a mouse designed to be highly susceptible to tumor formation. It was to aid in the development of anticancer therapies. A patent grants an exclusive right for a specified period of time to challenge the use of that technology or product by others without permission of the holder of the patent. Each country has its own set of rules governing the protection of intellectual property by patents or copyrights, although international agreements have established some common standards.

Patents are now allowed in the U.S. for genetically modified microbes, cells, animals, and plants, whether produced by recombinant DNA technology or by traditional breeding practices. New techniques for manipulating genes and inserting them into organisms can also be protected. A gene itself may be patented (the use of the gene and its sequence) but only if the function of the gene is known. This prevents patenting of unknown DNA sequences in hopes of figuring out what they are later. The rules governing patents in biotechnology evoked much controversy when they were first proposed, and they remain a complex subfield of patent law. While not a guarantee of exclusivity, patent protection attracted financial investment to the new biotechnology companies. This gave them the money and therefore the time to apply the new genetic technology to product development.

Even in the age of colossal multinational corporations, a technique, an idea, a few test tubes, some agar plates, and a patent pending can put a person into the genetic engineering business. Usually such operations begin in the university research laboratory of a faculty member but more than a few have started in a garage or vacant warehouse. The uses of world biotechnology grew exponentially from 1969 to 1984; since 1983, the growth has simply been explosive. The U.S. has tended to dominate the distribution of biotech companies in the world, partly due to the availability of investment capital and technical expertise, and partly due to favorable intellectual property laws. In 1988, there were 1,173 biotechnology companies worldwide, and 469 of them were in the United States (Table 4.4).

Most biotechnology companies in the U.S. are small, with 75% of the companies employing fewer than 50 people and fewer than 3% having more than 300 workers. They tend to be concentrated in metropolitan areas, as shown in Table 4.5 for the year 1994.

Nearly 1,300 biotechnology firms in the U.S. recorded an aggregate of over $7 billion in sales in 1993. The number of companies stabilized around 1,100 biotechnology firms, with nearly 200,000 employees, by the end of 1997. These employment figures do not include those working in government laboratories, academic scientists, institutes, or biotechnology centers. Therapeutics and health care account for the focus of about 55% of the companies and represent 65% of the employees (Dibner, 1998).

This late 20th century "cottage industry" spawned a new generation of biotechnology investment capitalists to provide the hundreds of millions of dollars of financial support needed to turn a golden idea or technology into a product. In 1992, compared

## TABLE 4.4
### Worldwide Distribution of Biotechnology Companies (1988)

| | |
|---|---|
| Developed nations | |
| United States | 469 |
| United Kingdom | 305 |
| Japan | 92 |
| Germany | 22 |
| Italy | 22 |
| France | 20 |
| Belgium, Denmark, Ireland, Netherlands, Sweden, Spain | 185 |
| Eastern Europe | 38 |
| | |
| Developing nations | |
| Taiwan | 11 |
| India | 2 |
| South Korea | 2 |
| Mexico, Brazil, China, Egypt, Pakistan | 1 each |
| | |
| Total | 1173 |

Source: *Biotechnology in Latin America. Politics, Impacts, and Risks*, Peritore, N.P. 1995.

## TABLE 4.5
### Location of U.S. Biotechnology Programs 1994

| | |
|---|---|
| San Francisco Bay | 181 |
| New York City | 136 |
| Boston | 128 |
| Washington, D.C. | 115 |
| San Diego | 100 |
| Los Angeles/Orange County | 68 |
| Texas | 52 |
| Seattle | 48 |
| North Carolina | 46 |

Source: Biotechnology Industry Organization, *Careers in Biotechnology* (pamphlet).

with total sales of $5.9 billion, biotechnology firms spent $4.9 billion on research, roughly equivalent to IBM's research and development budget that year. Like many new businesses, over 90% of these new ventures failed—most within the first five years, because they didn't convert their ideas into saleable products fast enough. Only 18% of 225 public biotechnology firms were profitable in 1995. Good scientists frequently are not good business managers.

The 1990s brought a slowdown in biotechnology investing as the fantastic forecasts failed to materialize. Nevertheless, by persevering or by shrewd alliances with established industries some successful biotechnology-based companies have emerged. Most are connected with health care, although agricultural applications continue to expand. Major corporations have also invested or acquired interests in biotechnology as they recognized the need for such processes or investments (Table 4.6).

### TABLE 4.6
#### Some Major Corporations Investing in Biotechnology

| | |
|---|---|
| Abbott Laboratories | Key Pharmaceuticals |
| Allied Chemical Corp. | Kimberly-Clarke |
| Allied Signal, Inc. | Life Technologies, Inc. |
| American Cyanamide Co. | Litton Bionetics, Inc. |
| American Home Products | Lubrizol Enterprises |
| American Hospital Supply Corp. | Merck and Company, Inc. |
| Amoco Corp. | Miles Laboratories, Inc. |
| Ares-Serono Laboratories | Miller Brewing Company |
| Baxter Travenol Labs, Inc. | Monsanto Agriculture Company |
| Becton Dickinson and Co. | National Distillers & Chemical Corp. |
| Bio-Rad Laboratories | New England Nuclear Corp. |
| Boehringer-Ingleheim Corp. | Novartis |
| Boehringer-Mannheim Corp. | Olin Corp. |
| Bristol-Meyers | Ortho Pharmaceutical Corp. |
| Campbell Soup Co. | Pennwalt Corp. |
| Celanese Research Co. | Pfizer, Inc. |
| Corning Glassworks | Phillips Petroleum Co. |
| Del Monte USA | Pioneer-Hi-Bred International, Inc. |
| Diamond Shamrock Biotechnology Research | Proctor and Gamble Co. |
| Dow Chemical Co. | RJR Nabisco, Inc. |
| E. I. du Pont de Nemours Co. | Rohm & Haas Co. |
| Eastman Kodak Co. | Rorer Group Inc. |
| Ecogen Inc. | Schering-Plough Corp. |
| Eli Lilly & Co. | Smith, Kline & French Labs |
| Exxon | Squibb Corp. |
| FMC Corp. | The Standard Oil Co. |
| General Electric Co. | Texaco Research Center |
| General Foods Corp. | 3M |
| Gist-Brocades USA, Inc. | Universal Foods Corp. |
| Glaxo-Wellcome Co. | The Upjohn Co. |
| Hercules Research & Development | W.R. Grace & Co. |
| Hoffman-LaRoche Inc. | Warner-Lambert Co. |
| International Mineral & Chemical Corp. | Weyerhauser Co. |
| Johnson & Johnson | Wyeth Laboratories |

*Source:* Brown, Sheldon S. *Opportunities in Biotechnology Careers,* pp.138–144.

The individual research and development expenditures of biotechnology companies extended into the hundreds of millions of dollars in 1995 (Table 4.7). Note that some companies have a large worth or market capitalization with relatively low research and development expenditures. This type of financial success depends both on the type of products and on the company's financial situation.

**TABLE 4.7**
**Top 25 Biotechnology Companies[a] Worldwide**
**by 1995 Research & Development Expenditures**

|  | Millions of $U.S. | Market Capitalization |
|---|---|---|
| 1. Amgen | $451.70 | 1. 16,255 |
| 2. Genentech | $345.30 | 2. 4,435 |
| 3. Chiron | $343.20 | 3. 3,312 |
| 4. Genetics Institute | $122.40 | |
| 5. Biogen | $87.45 | 5. 2,685 |
| 6. Immunex | $83.46 | 7. 1,124 |
| 7. Genzyme | $83.06 | 6. 1,778 |
| 8. Cephalon | $73.99 | 11. 491 |
| 9. Therapeutic Discovery | $68.92 | |
| 10. Centocor | $66.24 | 4. 2,157 |
| 11. British Biotechnology | $57.88 | |
| 12. SyStemix | $46.73 | |
| 13. Ligand | $41.64 | |
| 14. Vertex | $41.51 | 8. 830 |
| 15. NeXstar | $39.66 | |
| 16. Amylin | $39.34 | |
| 17. Gensia | $38.76 | |
| 18. Cor | $37.23 | |
| 19. Regeneron | $36.32 | 22. 186 |
| 20. Agouron | $36.32 | |
| 21. Alliance | $35.06 | |
| 22. Athena Neuroscience | $34.16 | |
| 23. RepliGen | $31.01 | |
| 24. Gilead | $30.36 | |
| 25. Liposome Co. | $30.15 | |

[a] Companies with publicly traded stock
Source: Revised from *Nature Biotechnology,* August 1996 (934–935) and *Genetic Engineering News,* June 1, 1997 (23).

Total spending of pharmaceutical and biotechnology companies on research and development, not all necessarily related to genetic engineering, continues to climb (Table 4.8).

### Table 4.8. Spending by Pharmaceutical and Biotechnology Companies on Research and Development

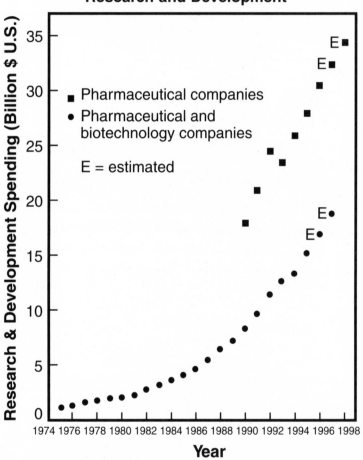

*Source:* Revised from Pharmaceutical Manufacturers Association Annual Survey, March, 1997, in: *Parexel's Pharmaceutical R&D Statistical Sourcebook.* Waltham, MA: Parexel International Corp., 1997, p.1.

Over the period from 1981 to 1992, a national survey showed that research and development expenditures in several categories (motor vehicles, office and computing, and food, beverages, and tobacco) leveled off while spending on medicines and drugs (including biotechnology) nearly tripled (Table 4.9).

It is becoming increasingly difficult to separate biotechnology and the pharmaceutical sector, because recombinant DNA technology is becoming so pervasive in that industry. The United States accounts for 36% of the worldwide company-financed research and development in pharmaceuticals (Table 4.10).

## Table 4.9. Manufacturing Industry R&D Performance 1981–1992

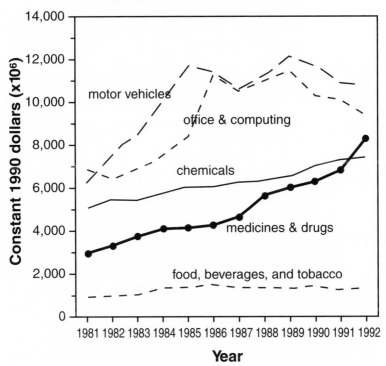

Source: Revised from National Science Board. *Science and Engineering Indicators—1996*. Washington, DC: U.S. Government Printing Office, p.269.

**TABLE 4.10**
Fraction of Worldwide Company-financed Pharmaceutical
Research and Development Ranked by Country for 1995

| | |
|---|---|
| United States | 36% |
| Japan | 19% |
| Germany | 10% |
| France | 9% |
| United Kingdom | 7% |
| Switzerland | 5% |
| Sweden | 3% |
| Italy | 3% |
| Other countries | 8% |

Source: Parexel's Pharmaceutical R & D Statistical Sourcebook, Waltham, MA: Parexel International Corp., 1997, p. 167.

These figures may actually underestimate the investment of the industrial sector as it is difficult to separate private industry-government coalitions in such countries as Japan and Cuba in which the economic model differs from that of the United States. The U.S. federal government investment in biotechnology research is considerable: $4.299 billion in fiscal year (FY) 1994, of which the lion's share (41%) was devoted to improving medicine and health-related programs. Much of the funding goes to universities, where it provides the basic groundwork to advance scientific knowledge and to stimulate application of new techniques through research foundations (39%), as well as supporting technology infrastructure in general (8%). Agriculture, manufacturing, environment, and energy resources receive much less governmental support (a total of 12%). In a world in which food and energy supplies dwindle as population and demand increases, some people believe that the social impact of genetic engineering research in these fields is potentially as dramatic as that of health care, if not more so. Research in several of these areas, particularly agriculture and manufacturing, is dominated by private industry in the U.S. (Table 4.11).

## Medical Applications

The top 25 publicly traded biotechnology companies worldwide (Table 4.7) are heavily invested in development of medical and health products. This position is driven by the ready application of genetic engineering to the production of scarce biomolecules, the

TABLE 4.11
# Federal Investment in Biotechnology Research Fiscal Year 1994 ($4.299 G)

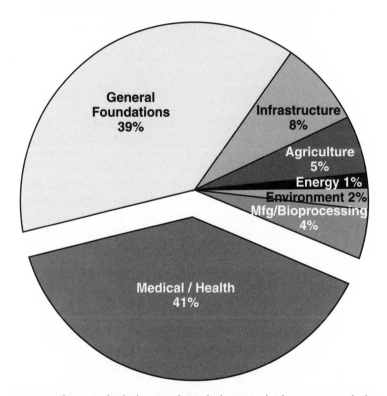

General Foundations 39%

Infrastructure 8%

Agriculture 5%

Energy 1%

Environment 2%

Mfg/Bioprocessing 4%

Medical / Health 41%

Source: National Science and Technology Council, Biotechnology Research Subcommittee, *Biotechnology for the 21st Century: A Report from the Biotechnology Research Subcommittee, Committee on Fundamental Science,* Washington, DC: Biotechnology Research Subcommittee, 1995, p.2.

relatively high rate of return on investment for pharmaceuticals, and a low profile on environmental impact since the recombinant organisms are controlled entirely within the production facilities. Those companies diversifying into other areas such as agriculture and bioremediation have encountered the ecological issues surrounding release of engineered organisms into the environment.

Recombinant biopharmaceutical agents are proteins that avoid immunological reaction by using human sequences of amino acids, based on human sequences of bases in the DNA. They are mostly hormones or growth factors, replacement enzymes (such as blood clotting factor VIII), or antibodies designed to neutralize or to target cell types for cancer therapy. The first biopharmaceutical agent produced with recombinant DNA technology was human insulin (Humulin®) from Eli Lilly. It was approved by the Food and Drug Administration (FDA) in 1982. The top selling biotech drug worldwide is Amgen's erythropoetin, an activator of blood cell production, ranking 15th in worldwide drug sales overall for 1996. Four recombinantly engineered drugs were among the top 50 drugs sold in 1997. To put it in perspective, though, biopharmaceuticals accounted for only about 10% of U.S. drug sales in 1996 (Dibner 1998).

Gene therapies for cancer, cystic fibrosis, and lysosomal storage diseases (by enzyme replacement) are also being developed. Whereas W. F. Anderson's proof of concept for gene therapy in 1990 related to a rare disease (replacement of the enzyme adenosine deaminase (ADA) in ADA-deficient immunocompromised individuals), the diseases now being targeted afflict significant numbers of people. For both technical and ethical reasons, no permanent germline genetic repairs are being advanced to replace the defective gene or otherwise inactivate it in the reproductive cells, which would transmit the modification to future generations. Only the individual being treated (presumably) will be affected by the therapy. Targeting and the robustness and duration of the therapeutic effect will be carefully monitored. The difficulties of this approach are legendary; the ability of such treatments to compete with more conventional pharmacologic avenues, when they can be developed, will decide the success of these companies.

A concern of both proponents and opponents of gene therapy is the risk of abuse of the technology. While there is general agreement that gene therapy should be subject to societal controls, defining what constitutes abuse will be difficult. The NIH Recombinant Advisory Committee, charged with approving gene therapy protocols, is holding a series of conferences and open forums designed to bring the public into the debate on what sorts of treatments are medically justified. A member of this committee, bioethicist Eric Juengst from Case Western Reserve University in Cleveland, Ohio, observed "Just about any enhancement could be packaged as therapy in some context." His thoughts are echoed by Sheila Rothman of Columbia University in New York, who says "Powerful forces are available to promote enhancement, and medical treatments can quickly become enhancements." A hypothetical case in some discussions: would a gene therapy solution to male pattern baldness be a "therapy" or an "enhancement"?

# Agriculture and Manufacturing

Agricultural products register the largest commercial impact outside of medically related uses of genetic engineering. In 1992, out of a total of $5.9 billion in receipts, agribiotech accounted for sales of $184.5 million. Large chemical and pharmaceutical companies such as Monsanto, Dupont, and G.D. Searle have diversified by investing in agricultural applications of genetic engineering through interests in other companies like Collagen, Biogen, Genentech, Genex, and Biotechnica International (see Table 4.6). These companies deal in proprietary technologies rather than the commodities themselves: They provide the engineered seed and supporting treatments such as herbicides. Although traditional scientific breeding of plants has flourished for hundreds of years, developing desired qualities by *in vitro* propagation of plants from clones derived from single cells has been a twentieth century phenomenon. Genetic engineering, in the form practiced at the molecular level for bacteria, fungi, and animal cells, was slowed by the initial lack of suitable vector systems for transfecting plant cells and by difficulty in penetrating the tough plant cell wall. Some types of crops such as rice, bananas, and cereals that would impact food production for the worker population in developing nations have remained recalcitrant to molecular manipulation. The full-scale use of engineered plants, unlike most recombinant bacterial or fungal applications, necessarily involves release of large numbers of recombinant organisms over millions of acres of cropland. Although they cannot walk, swim, or fly away, plants can disperse their genetic material (pollen and seeds) over considerable distances, potentially interbreeding with related weed plants. Pest resistance to specific recombinant strategies can also develop, conceivably causing disastrous crop failures when pest populations surge. Environmentally tolerant plants could move into new ecosystems disrupting the indigenous species. As a result, the ecologic consequences of agriculturally based genetic engineering have attracted much attention and have been a prime target for protest groups and governmental regulation. Public misgivings about pest management by genetic engineering have drastically slowed deployment of the technology. Dr. R. Jones Cook of the U.S. Agriculture Research Service at Washington State University, Pullman, Washington, reflects the current climate in his statement, "[the] question of social acceptance of pest management with transgenes...remains a deterrent to plant biotechnology" (American Association for the Advancement of Science, 1998).

The approach to regulating genetic engineering in the United States has been to expand the auspices of existing regulatory agencies and to provide supplemental legislation. The result is a complex network of interconnected responsibilities. Table 4.12 outlines the regulatory responsibilities for reviewing planned introductions of genetically modified organisms, including plants.

**TABLE 4.12**

**Responsibilities for Reviewing Planned Introductions of Genetically Modified Organisms**

| | |
|---|---|
| No release in environment (contained) | |
|     Federally funded | Funding agency[a] |
|     Nonfederally funded | NIH or S&E voluntary review, APHIS[b] |
| | |
| Foods/food additives, human drugs, | |
| medical devices, biologics, animal drugs | |
|     Federally funded | FDA[d], NIH guidelines and review |
|     Nonfederally funded | FDA[d], NIH voluntary review |
| | |
| Plants, animals, animal biologics | |
|     Federally funded | Funding agency[a], APHIS[b] |
|     Nonfederally funded | APHIS[b], S&E voluntary review |
| | |
| Pesticide organisms | |
|     Genetically engineered: | |
|         Intergeneric | EPA[c], APHIS[b], S&E voluntary review |
|         Pathogenic intrageneric | EPA[c], APHIS[b], S&E voluntary review |
|         Intrageneric nonpathogen | EPA[c], S&E voluntary review |
|     Nonengineered: | |
|         Nonindigenous pathogens | EPA[c], APHIS[b] |
|         Indigenous pathogens | EPA[c], APHIS[b] |
|         Nonindigenous nonpathogens | EPA[c] |
| | |
| Other uses (microorganisms) released in | |
| the environment | |
|     Genetically engineered: | |
|         Intergeneric organisms | |
|             Federally funded | Funding agency[a], APHIS[b], EPA[c] |
|             Commercially funded | EPA, APHIS, S&E voluntary review |
|         Intrageneric organisms | |
|             Pathogenic source organisms | |
|                 Federally funded | Funding agency[a], APHIS[b], EPA[c] |
|                 Commercially funded | APHIS[b], EPA[c] (if nonagriculture use) |

**TABLE 4.12** *continued*

| | |
|---|---|
| Intrageneric combination | |
| Nonpathogenic source organisms | EPA Report |
| Nonengineered organisms | EPA Report*, APHIS[b] |

* Lead agency
[a] Review and approval of research protocols conducted by NIH, S&E, or NSF
[b] APHIS = Animal and Plant Health Inspection Service (involved when microorganism is a plant or animal pathogen or regulated by permit)
[c] EPA = Environmental Protection Agency (jurisdiction for plots of more than 10 acres)
[d] FDA = Food and Drug Administration (reviews federally funded environmental reasearch only when it is for commercial purposes)
NIH = National Institutes of Health
S&E = U.S. Department of Agriculture, Science, and Education

*Source:* 51 Fed. Reg. 23305

These regulations cover basic academic research and commercial releases. Commercial biotechnology products also have to be approved by the appropriate government agency (or agencies), depending on the intended use of the product (Table 4.13).

**TABLE 4.13**
**U.S. Federal Government Agencies Responsibilities for Approval of Commercial Biotechnology Products**

| | |
|---|---|
| Food/food additives | FDA[a], FSIS[b] |
| Human drugs, medical devices and biological products | FDA |
| Animal drugs | FDA |
| Animal biological products | APHIS |
| Other contained uses | EPA |
| Plants and animals | APHIS[a], FSIS[b], FDA[c] |
| Pesticide organisms released into environment (all) | EPA[a], APHIS[d] |
| Other uses (microorganisms) | |
| 1. Intergenera(ic) combination | EPA[a], APHIS[d] |
| 2. Intragenera(ic) combination: | |
| (a) pathogenic source organism | |
| agricultural use | APHIS |
| nonagriculture use | EPA[a,e], APHIS[d] |
| (b) no pathogenic source organisms | EPA Report |
| nonengineered pathogens | |
| (i) agriculture use | APHIS |
| (ii) nonagriculture use | EPA[e], APHIS[d] |
| nonengineered nonpathogens | EPA Report |

[a] Lead agency (FDA =Food and Drug Administration; EPA = Environmental Protection Agency)
[b] FSIS = Food Safety and Inspection Service (under the Asst. Secretary of Agriculture for Marketing and Inspection Service)
[c] FDA is involved in relation to food use
[d] APHIS = Animal and Plant Health Inspection Service (involved when microorganism is a plant or animal pathogen or regulated by permit).
[e] Only for significant new use (proposed new rule, 1989)

*Source:* 51 Federal Register 2339

Over the past 10 years more than 3,600 field trials (releases) have been approved by the Animal and Plant Health Inspection Service of the United States Department of Agriculture (USDA) (Table 4.14).

## Table 4.14
# Categories of Field Releases 1987–1997

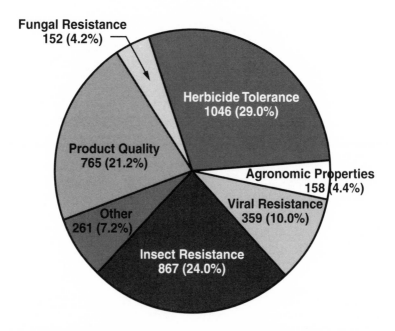

*Source:* Revised from Animal and Plant Health Inspection Service (APHIS), BSS Biotechnology Update (November, 1997); http://www.aphis.usda.gov/bbep/bp/newsletter.html.

From the time of submission of a formal proposal for field release of genetically modified organisms, the Department of Agriculture has 120 days to review and approve the proposal. After a number of years of testing with no evidence of ecological impact, companies working with genetically engineered plants can request that APHIS determine that there is "no potential for plant pest risk" and certify the release of the tested varieties.

Table 4.15 reveals that over a two year-period a number of engineered plants that had initially been ecologically suspect in principle showed no evidence under the particular conditions in which they were tested that they would become "weedy" or transfer their modified genetic material to native plants. While this doesn't prove that they are safe in all environments, the initial apprehension turned out not to be well founded.

**TABLE 4.15**
**No Potential for Plant Pest Risk Determinations (APHIS) 1996–1997[a]**

| Product | Company | Date |
|---|---|---|
| Insect-resistant corn | Northrup King Company | January 1996 |
| Herbicide-resistant cotton | Dupont Agricultural Products | January 1996 |
| Male-sterile corn | Plant Genetic Systems(America), Inc. | February 1996 |
| Altered ripening tomato | Agritope, Inc. | March 1996 |
| Colorado potato beetle–resistant potato | Monsanto Agricultural Company | May 1996 |
| Virus-resistant squash | Asgrow Seeds | June 1996 |
| Herbicide-tolerant soybean | AgroEvo | July 1996 |
| Virus-resistant papaya | Cornell University & University of Hawaii | September 1996 |
| Insect-resistant corn | DeKalb Genetics Corporation | March, 1997 |
| Insect-resistant cotton | Calgene, Inc. | April 1997 |
| High-oleic acid soybean | Dupont Agricultural Products | May 1997 |
| Insect-resistant corn | Monsanto Agricultural Company | May 1997 |

[a] APHIS = Animal and Plant Health Inspection Service. Potential for plant pest risk is a measure of the chance that the genetically altered organism will cross-breed with closely related wild relatives to produce a weed plant with the new characteristic and significant growth or propagation advantage.

*Source:* Data from APHIS BSS Biotechnology Update (November 1997), U.S. Department of Agriculture (website: http://www.aphis.usda.gov)

134 Facts, Data, and Opinions

Rather than writing new regulations for genetically engineered products, the U.S. Government has extended coverage
with the same legal statutes that already protect agriculture and
the nation's food supply from pests and contamination from conventional sources. This policy decision has been criticized by
those who would prefer genetically engineered organisms and
produce to be singled out for special regulation. The statutes and
the U.S. Legal Code are a very large document that has been revised and amended over the years. It is most readily accessed
and searched via the Internet (http://www.law.cornell.edu/us-
code/index.html and other sites). The indicated sections of the
following statutes (Title | U.S. Code | Section; Title 7 = Agricul-
ture, Title 21 = Food and Drugs) are considered applicable to
USDA-regulated biotechnology (51 Federal Register, Office of
Science and Technology Policy Federal Registration Notice on a
Coordinated Framework for Regulation of Biotechnology, 1986,
p. 23339), but have no special mention of biotechnology in the
documents: Virus-Serum Act of 1913 (21 U.S.C. 151–158), Federal
Plant Pest Act of May 23, 1957 (7 U.S.C. 150aa–150jj), Plant Quar-
antine Act of August 20, 1912 (7 U.S.C. 151–164, 166, 167), Or-
ganic Act of September 21, 1944 (7 U.S.C. 147a), Federal Noxious
Weed Act of 1974 (7 U.S.C. 2801 et seq.), Federal Seed Act (7
U.S.C. 155 et seq.), Plant Variety Protection Act of 1930 and 1970
(7 U.S.C. 2321 et seq.), Federal Meat Inspection Act of 1907 (21
U.S.C. 601 et seq.), and The Poultry Products Inspection Act of
1957 (21 U.S.C. 451 et seq.).

This is not an exhaustive list of applicable statutes. Modifications are made as technology and new issues develop. Keeping
up with constantly changing biotechnology regulation is complex and time consuming. The BioTechnology Permits Home
Page at http://www.aphis.usda.gov/BBEP/BP/ is a useful
source for information on regulations pertaining to agricultural
biotechnology and links to other governmental agencies and
their regulations.

Genetic engineering of foodstuffs is an even more complex
and emotionally involved issue than the environment. According to University of Wisconsin sociologist Frederick Buttel
(Thompson, 1997), food biotechnology is responsible for 90% of
the controversy while medical biotechnology accounts for 90%
of the products of genetic engineering. The human cloning controversy has temporarily shifted the balance of discussion since
that statement was made. Debate over the use of recombinant
DNA technologies in the production of food includes not only

environmental and consumer safety issues but the right to choose whether one uses a product produced in a certain way ("synthetic" vs. "natural"). In response to this debate, two bills were introduced into the U.S. House of Representatives (H.R. 2084 and H.R. 2085). They would require the distinct labeling of milk and milk products derived from cows treated with synthetic bovine growth hormone to stimulate milk and meat production intended for human consumption and the development of technologies to verify the source of the bovine hormone in foods. H.R. 2084, the "Bovine Growth Hormone Milk Act," imposes labeling requirements for milk and milk products produced from cows treated with synthetic bovine growth hormone and directs the development of a test specific for the synthetic hormone. It also requires that the Department of Agriculture pay a lower subsidy price for milk and milk products produced with the aid of synthetic growth hormone. H.R. 2085, the "Bovine Growth Hormone Milk Act," imposes the same requirements as H.R. 2084 without requiring the Department of Agriculture to pay a lower price for milk and milk products produced with the aid of synthetic growth hormone. Both were introduced by Mr. Sanders of Hawaii (and 27 other representatives) on July 20, 1997, and were referred to the Committee on Agriculture.

Despite the confusing web of regulation, the perceived future returns on investment of genetically engineered crops have kept entrepreneurial interest high. As shown in Table 4.16, in addition to the expected agricultural and food-based players, large chemical and pharmaceutical companies lead as the most active applicants to the USDA for field testing of transgenic crops.

The EPA has indicated its intention to classify as pesticides all transgenic plants that are disease- or insect-resistant. Biotechnology companies have protested the expense and delay engendered by the legal and bureaucratic process of pesticide regulation employed for chemical pesticides. They point out that the same resistance gene introduced by traditional breeding would not be considered a pesticide and would not be regulated. Small companies with a niche product lacking the high-powered regulatory machinery of the large integrated chemical/agricultural companies feel that they are being unfairly excluded from the marketplace. The list of genetically engineered agricultural products presently on the market is short but likely to grow substantially (Table 4.17).

Most of these products promise improved production through pest-, herbicide-, and pesticide-resistance, although

**TABLE 4.16**
Most Active Applicants to USDA for Field Testing of Transgenic Crops

| Applicants | % of Applications[a] |
|---|---|
| Chemical companies | 46% |
| Monsanto (52%) | |
| Upjohn (Asgrow Seed) | |
| Dupont | |
| Sandoz (Northrup, King, and Rogers NK Seed) | |
| Ciba-Geigy | |
| Hoechst-Roussel | |
| Imperial Chemical Industries | |
| American Cyanamid | |
| Universities / USDA | 17% |
| USDA Agriculture Research Service | |
| Cornell University | |
| North Carolina State University | |
| University of Kentucky | |
| University of California | |
| Michigan State University | |
| Seed Companies | 15% |
| Pioneer Hi-Bred (51%) | |
| DeKalb Plant Genetics | |
| Holden's Foundation Seed | |
| Petoseed | |
| Harris Moran | |
| Biotechnology Companies (only biotech) | 13% |
| Calgene (82%) | |
| DNA Plant Technology | |
| Agrigenetics | |
| Food Companies | 5% |
| Frito-Lay | |
| Campbell | |
| Heinz | |
| Land O'Lakes | |
| Miscellaneous | 4% |
| Cargill | |
| Amoco Technology | |

[a]Compilation of 549 applications to U.S. Department of Agriculture (USDA) from 1987 to 1993.

*Source:* Union of Concerned Scientists report, 1993.

**TABLE 4.17**
**Genetically Engineered Agricultural Products on the Market**

| Product | Company | Enhancement |
|---|---|---|
| Herbicide-resistant: | | |
| Liberty Link™ Corn<br>Liberty Link™ Canola | AgroEvo (Canada) | Liberty™ herbicide-resistant<br>Liberty™ herbicide-resistant |
| IMI-CORN®<br>IMI™-Canola Seed | Pioneer H-Bred Intnl.<br>(Des Moines, IA) | Imidazolinone herbicide-resistant<br>Imidazolinone herbicide-resistant |
| BXN® Cotton | Calgene, Inc. (Davis, CA) | Herbicide-resistant |
| DEKALB GR<br>Hybrid Corn | DeKalb Genetics Corp.<br>(DeKalb, IL) | Glufosinate herbicide resistance |
| Pest-resistant: | | |
| DEKALBt™ Insect-<br>Protected Corn | DeKalb Genetics Corp.<br>(DeKalb, IL) | Corn borer-resistance |
| Maximizer™ Hybrid Corn | Novartis Seeds<br>(Minneapolis, MN) | Corn borer-resistance |
| Freedom II™ Squash | Seminis Vegetable Seeds<br>(Saticoy, CA) | Plant virus-resistant |
| Gray Leaf Spot–Resistant Corn | Garst Seed Co. (Slater, IA) | Disease-resistant |
| Stress-resistant: | | |
| High pH Tolerant Corn | Garst Seed Co. (Slater, IA) | Alkaline soil–tolerant |
| Augmented production: | | |
| Optimum® Soybeans<br>Products, (Wilmington, DE) | Dupont Agricultural | High oleate oil content |
| Laurical® | Calgene, Inc. (Davis, CA) | Special manufacturing food oil |

*Source:* Revised from Biotechnology Industry Organization data reported in *Genetic Engineering News* Vol. 17 (12), June 15, 1997.

agricultural products with genetically altered oil composition are available and an even more diverse set of products is in the development pipeline. Published Premanufacture EPA Notifications include a number of applications involving contained fermentation as well as release into the environment that are directed at production of manufacturing intermediates or products.

Microorganisms are responsible for a large number of commercial products generated through conventional biotechnology. Natural products, or secondary metabolites, are molecules, often chemically complex, that are synthesized by an organism to gain a selective advantage for survival and growth in a particular environment. Some, such as the bread mold organism, are the ancestors of modern pharmacological treatment and have been enshrined in folk medicine by indigenous human societies for treatment of various ailments. The most recognizable of these are the antibiotics, which have apparently evolved to reduce competition by other microbes for space and nutrients. The isolation and clinical use of antibiotics as medicines began with penicillin, discovered in 1928 by English bacteriologist Sir Alexander Fleming. Along with great strides in improving public sanitation, antibiotics have been credited for the reduction in disease-related deaths in developed and developing nations since 1900.

Many pharmaceutically useful agents and novel chemical structures are produced by microorganisms. Other materials including food supplements and emulsifying agents for stabilizing fat-containing mixtures are obtained through microbial fermentation. Besides providing scientists with the tools to design new generations of antibiotics to overcome evolving drug resistance, recombinant DNA techniques are used to move the genes responsible for natural product synthesis into organisms more amenable to and safer to culture.

In the quest for decreased reliance on petroleum-derived precursor chemicals, a number of industrially useful building blocks such as glycerol and acetone have been produced at competitive cost by microorganisms cultured on low-grade carbohydrate residue from processed agricultural plant waste. Polymeric substances produced by microorganisms, such as the food additives xanthan gum and various alginates and the biosynthetic "plastic," polyhydroxybutyric acid (from *Alcaigenes eutrophus*), are available from microorganisms. Japanese scientists at research institutes in Tsukuba and Kyoto, Japan, reported in 1997 having engineered a cyanobacterium (*Synechococcus* species) to produce up to 10% of its dry weight as polyhydroxybutyric acid using only light energy and carbon dioxide. Through genetic engineering, the agricultural company Agracetus is constructing cotton plants that will produce fibers of cotton (cellulose) filled with a synthetic polyester fiber—a natural wash-and-wear fiber. Dupont (Wilmington, Delaware) and Genecor International, Inc. (Palo Alto, California) have incorporated genes for bacterial and yeast enzymes in a single organism to make 3G (trimethylene glycol), the monomer base of a new recyclable polyester (3GT). The bottom line remains the economic feasibility of production from renewable biological sources rather than limited petrochemical stores.

The use of transgenic sheep and goats to produce therapeutic products in their milk is now well established. "Pharming" of human proteins can be done on a large, formerly industrial scale, with few effects on the animals. A mature rabbit yields up to 8 liters of milk per year, a sheep 300 liters, a goat 1,000 liters, and a cow 8,000 liters. Thus, small herds of these animals producing at expression levels of 8 to 15 grams of product per liter could be quite competitive with industrial-scale cell culture of complex proteins, and quite economical.

Other products in various stages of clinical trials include human anti-thrombin III (ATIII), produced in goats for acquired ATIII deficiency, which is a blood coagulation proteolytic disorder (Genzyme Transgenics Corp., Framingham, Massachusetts), $\alpha$1-antitrypsin, a protease inhibitor produced in sheep for treatment of cystic fibrosis (PPL Therapeutics, Edinburgh, Scotland), and $\alpha$-glucosidase, a carbohydrate-hydrolyzing enzyme, produced in rabbits for the treatment of glycogen storage disease type II, which results from a deficiency of this enzyme (Pharming Healthcare Products, Leiden, The Netherlands). Other proteins are under development, including human serum albumin used as a blood volume extender for surgery and trauma. Currently, 440 metric tons of the protein, extracted from 16 million liters of human blood plasma, are required for therapy worldwide each year. Transgenic animal production of human proteins would avoid issues of hepatitis, HIV, Creutzfeld-Jacob prion, and other pathogen contamination passed on from human blood supplies—although the potential for transmission of animal diseases across species does need to be addressed. However, recent governmental action in Europe places the future of this industry in doubt. In March, 1998, the Dutch Ministry of Agriculture, partly in reaction to the nuclear transplant cloning controversy, reversed an earlier decision and revoked Pharming Healthcare

Products' permit for nuclear transplant-based transgenic mammal cloning work designed to speed the generation of transgenic animals producing human therapeutic proteins. This judgment effectively bans nuclear transplant cloning in the Netherlands for the foreseeable future.

# Bioremediation

As the word's component parts suggest, bioremediation means to repair or make right using biological processes. The word is most commonly employed to refer to the removal of manmade contaminants from the environment. In fact, the process is neither new nor restricted to human contamination. Recycling of nutrients, minerals, and chemical building blocks has proceeded for millions of years to make way for new life, and fortunately so, otherwise the earth would be deeply buried under dead bacteria, insect exoskeletons, and leaf litter. With the increasing pressure of burgeoning human populations and the impact of industrialization, humankind has had to find ways to speed up recycling and to clean up concentrations of toxic substances from commercial, governmental, and residential waste and to cope with accidental spills. The difficulty of extracting toxic materials present at the part per million level (roughly one drop in 13 gallons) and distributed throughout large volumes of earth, water, or air has led to some innovative treatment strategies. Studies employing naturally occurring microorganisms and plants that use these toxins as food sources or that concentrate and store heavy metal ions have shown that under some circumstances the desired results can be achieved. Genetic engineering can potentially combine several steps of a metabolic pathway in a single microorganism. Its phenotype could be structured to include traits that would not allow the organism to survive outside of the treatment zone, such as requiring the contaminant as a sole carbon source. Fewer types of organisms would then need to be present to degrade many types of contaminants more completely and the possibility of escape of the altered microbes into the natural bacterial population would be reduced. The widely distributed root systems of some plants can take up toxic metals and with the help of recombinant DNA technology can achieve higher specificity and greater concentration of the metals from both land and water.

So why aren't these environmental cleanup tools in widespread use? The National Priority List of toxic waste sites contains more than 1,200 confirmed sites, with perhaps more than 32,000 sites potentially needing remediation and more being added to the list yearly. In addition, a significant number of the 7 million underground storage tanks in the U.S. are leaking, all of which points to an impending crisis of contamination when these tanks reach and pass their designed life span.

With their insatiable capacity to multiply, microorganisms have extended their domain into environments well beyond that which is normally considered conducive to life. From oil reservoirs and ore deposits buried thousands of feet below the earth's surface to boiling hot springs, to superheated acid underwater volcanic jets, these single-celled organisms live and reproduce. They thrive on chemicals that we consider toxic waste—sulfuric and phosphoric acid, or carcinogen-laced oils and tars, breaking them down into carbon dioxide, water, and mineral elements. The energy and chemical building blocks they extract in this so-called mineralization process are used to make new organisms and to erect barriers against their caustic environment. Soil, water, and air can all be treated by bioremediation. On a familiar level, microorganisms process biological waste in sewage and septic systems all over the world. Microorganisms adapt to utilize the sources of energy and raw materials that are available to them. Those that can't are "outcompeted" for whatever food sources remain, and are eliminated or reduced to low levels in the microbial population. *Escherichia coli,* the common human intestinal bacterium, contains more than 4,300 protein-encoding genes. In a normal cell growing on glucose as a carbon and energy source, more than 800 enzymes are expressed to catalyze the intertwined cellular reactions we call life. Although *E. coli* are not particularly adept at using environmental contaminants as metabolic fuel, other species such as *Pseudomonas, Clostridium, Brevibacterium,* and *Desulfobacterium* readily adapt, taking advantage of various combinations of metabolic enzymes carried on extrachromosomal DNA plasmids (much like antibiotic resistance factors). Many chemical reactions are carried out: hydrolysis, hydroxylation, methylation, dealkylation, nitro reduction, deamination, ether cleavage, numerous conjugation reactions, and dehalogenation (dechlorination), all of which are needed to break down the toxic chemicals. This last process is particularly useful, since the difficult part of many bioremediation efforts is the removal of the highly stable chlorine atoms from industrial environmental pollutants such as DDT as well as from those listed in Table 4.18. Often one species of organism carries out one or only

**TABLE 4.18**
Examples of Environmental Contaminants

| Contaminant | Used in |
|---|---|
| Chlorobenzoic acids | Degradative products of polychlorinated biphenyls (PCBs), herbicides, and plant growth regulators |
| Chlorinated biphenyls | Transformer and hydraulic fluids, plasticizers, fire retardants |
| Chlorobenzenes | Industrial and paint solvents, by-products of textile dyeing, fungicides |
| Chlorophenols & derivatives | Antifungal agents, wood preservatives, herbicides (2,4-D; 2,4,5-T) |
| Phenylamide herbicides | Herbicides |
| Chlorinated dioxins and furans | Burning of PCBs, manufacturing byproducts, hydraulic and heat exchanger fluids |

*Source:* Baker, K.H., and D. S. Herson (eds). *Bioremediation.* New York: McGraw Hill, 1994, p. 49.

a few parts of the degradation pathway, so cooperation between species in the form of a mixed bacterial culture is generally required to completely degrade resistant compounds.

Eastern Europe is a major site for toxic waste cleanup efforts due to lax environmental controls on industry in that region since World War II, with an estimated 50,000 contaminated sites in Bulgaria alone. The toxic heavy metal ions copper, zinc, cadmium, arsenic, and even uranium that pollute soils can be immobilized by microbial conversion to insoluble forms that leach only minimally. Carcinogenic hydrocarbons and long-lived halogenated hydrocarbons are also being targeted for degradation by endogenous bacterial species in several studies. Much of the front line research is being done by eastern European scientists in Budapest and Prague, targeting sites such as the Kola Peninsula in Russia (heavy metals, sulfur dioxide), Northern Bohemia (strip mining), and Upper Silesia (industry and poor waste management) (Aldridge, 1997).

Biochemical reactions catalyzing the degradation of environmental contaminants can also be exploited to produce saleable products on their own. Organisms such as the naturally occurring *Thiobacillus ferrooxidans* and other more thermophilic (heat-loving) organisms release economically important metals such as gold and cobalt from their insoluble complexes in ores that are refractory to the normal industrial cyanide leaching process. Thus they make recovery of metals from low grade ores economically feasible. Biocatalytic removal of sulfur from petroleum and coal (desulfurization) using endogenous bacteria and oxygen to produce cleaner burning fuels is being pursued by companies such as Energies Biosystems Corporation (The Woodlands, Texas) to meet the refiners' growing need to decrease sulfur emissions from fossil fuels. Other biorefining technologies being considered include nitrogen and metal ion removal, viscosity reduction, and cracking (hydrocarbon chain length reduction) to make products like gasoline, processes normally carried out in refineries at high temperature and pressure. The LATA Group (Ochelata, Oklahoma) provides selected water-soluble nutrients to stimulate indigenous beneficial microorganisms in petroleum fields and fuel storage facilities. The damaging bacteria that produce the highly corrosive hydrogen sulfide responsible for low petroleum yields and spoilage of petroleum are ecologically outcompeted, reducing the amount of sulfur contaminants in the fuel.

Genetic engineering offers the capability of combining several components of a chemical degradation pathway in a single organism. This organism may also be fitted with biological safeguards beyond natural competition when the foreign substrate is exhausted. Table 4.19 lists a number of types of bioremediation in which different contaminated environmental systems can be treated by altering the method of application of the appropriate microorganisms.

Numerous small companies have sprung up around bioremediation technologies, but the tendency is for them to merge or be taken over by larger organizations that may require their services. International Bioremediation Services (Walsall, U.K.), Microbe Masters of InterBio (Baton Rouge, Louisiana), PolyBac (Bethlehem, Pennsylvania), and Oppenheimer Biotechnology (Austin, Texas) are just a few of the survivors with their varied technologies. Since each bioremediation situation may require a substantially unique solution, a number of analysts support the formation of bioremediation consortia to broaden their credibility and strengthen their position on legal liability issues. The U.S. Department of Energy (Germantown, Maryland) faces the environmental cleanup of 2.5 trillion liters of contaminated liquids and some 200 million cubic meters of contaminated solids. The

### TABLE 4.19
#### Types of Bioremediation

| | |
|---|---|
| Bioaugmentation | Addition of bacterial cultures to medium |
| Biofiltration | Microbes immobilized on columns to treat air emissions |
| Biostimulation | Optimization of microbes already present in medium |
| Bioreaction | Biodegradation in a container or reactor |
| Bioventing | Oxygenating contaminated soils to stimulate microbial growth and activity |
| Composting | Aerobic thermophilic treatment, addition of bulking agent |
| Land forming | Solid phase treatment for contaminated soil in place or after removal |

Source: Revised from Baker, K. H., and D. S. Herson (eds). *Bioremediation.* New York: McGraw Hill, 1994, p. 3.

agency, which pursues more traditional methods of environmental restoration, also funds scientific research on alternative methodologies including bioremediation. In 1996 it established BASIC (Bioremediation and its Social Implications and Concerns), a program designed along the same lines as the Human Genome Project's ELSI (Ethical, Legal, and Social Implications) commission, which is similarly supported by the DOE along with the National Institutes of Health. Like ELSI, this bioremediation organization will have to bridge the often conflicting interests of the scientific researchers, public advocates for community involvement such as the Waste Policy Institute (Washington, D.C.), and government regulators. Preliminary findings from opinion surveys conducted by the governmental agencies Environmental Canada and Industry Canada indicate that "the public expects to be consulted in establishing guidelines or codes of ethics for biotechnology" (Fox, 1996), yet the public lacks specific knowledge about biotechnology and genetic engineering, especially in the area of the environment. Clear explanations will be required to avoid hasty and uninformed public reaction.

Plant-based bioremediation (phytoremediation) is applicable to some aspects of environmental cleanup, though presently it commands a significantly smaller market. The total 1997 U.S. market for phytoremediation is estimated to be $3 million to $7 million, compared to $200 million to $250 million for microbial remediation, according to D. Glass Associates, Inc., Needham, Massachusetts (Genetic Engineering News, 1997). By 2005, the figures are predicted to be more comparable: $100 million to $200 million for phytoremediation, versus $400 million to $700 million for micro-bial remediation. Marshes and estuaries have long been known to cleanse heavy metal ions and excess nutrients in fertilizer runoff from agricultural land. A small number of companies focusing en-tirely on phytoremediation are experimenting with amaranth, In-dian mustard, and sunflowers to absorb toxic or radioactive metal ions from contaminated soil and water with the assistance of both the EPA and Phytotech, Inc. (Monmouth, New Jersey). Field trials in Chernobyl, Ukraine, are being run to demonstrate cleaning of the dangerous radioactive elements, cesium and strontium, re-maining from the atomic reactor accident there. Crossbreeding and genetic engineering are being used to further refine and increase the concentration of these materials by these plants. They can then be safely disposed of or recycled from the harvested plants. Phy-toWorks, Inc. (Gladwyne, Pennyslvania) and EarthCare, Inc. (Hanover, New Hampshire) are also concentrating on toxic metal ions such as mercury and on organic chemical contaminants. Phy-todegradation of organic chemical contaminants in ground water is being pursued by Applied Natural Sciences (Hamilton, Ohio), Ecolotree (Iowa City, Iowa), and PhytoKinetics, Inc. (Logan, Utah). PhytoKinetics, Inc. is also looking at root-based detoxification of organic compounds in soil by grasses and other plants.

Genetic engineering can also enhance phytoremediation. Transgenic tobacco plants making engineered antibodies to atrazine, a widespread herbicide contaminant in soil and water, bind and concentrate the toxin while allowing the plant to sur-vive. Aberdeen University (Scotland, U.K.) and Axis Genetics (Cambridge, U.K.) plan to transfer the antibody genes to deep rooted plants such as rape and other Brassicas. This would be used as a final cleanup step after most of the atrazine is removed by microbial remediation.

Phytoremediation research is also being carried out at a number of U.S. universities including Cornell University, Iowa State University, Kansas State University, Montana State Univer-sity, Ohio University, Oklahoma State University, Rutgers Uni-versity, and the Universities of Georgia, Iowa, Maryland, Missouri, Oklahoma, and Washington. Heffield University and Glasgow University are involved overseas. Various U.S. govern-ment agencies sponsor extramurally (by giving grants) and/or conduct their own laboratory research on phytoremediation. These include the Department of Agriculture, the EPA, the Army Corps of Engineers, and the Department of Energy. Some non-profit organizations such as Argonne National Laboratories, Los

Alamos National Laboratories, the Institute of Gas Technology/Gas Research Institute, and the Rothamstead Experimental Station (United Kingdom) also have programs. A major impediment to the widespread use of bioremediation to remove environmental contaminants is the question of whether the genetically modified organisms released into the environment will escape control measures and adversely affect the ecosystem. Safety is, of course, a relative term that is defined by tolerability and acceptability limits. These are themselves set according to currently available expertise and, to the chagrin of proponents of environmental uses of genetically modified organisms, increasingly by public reaction. The other part of the evaluation process is risk assessment. This is a more quantitative measure of the probability of harm occurring, encompassing both the probability (chance) of its happening and the amount of damage incurred should it happen. In the case of modified plants or microbes decisions have been primarily based on safety since it is generally not possible at this time to conduct a straightforward risk assessment on organisms, whether genetically modified or not. Chemical toxicology, which has accumulated evidence through long experience, supplies information about classes of chemicals and their effects, the dosages required, and probabilities of adverse effects predicted by validated models as part of the risk assessment–risk management process. With plants, animals, and microbes, although similar organisms and modifications are used to compare with previously analyzed systems, dose-response effects cannot be rigorously determined, whether or not they are genetically modified. Robust models for living organisms remain to be developed. The particular hazards to be expected may not necessarily be predictable with organisms either. The current compromise for evaluating organisms requires less quantification than chemical risk assessment, but attempts instead to identify the hazards of greatest concern. An example of regulatory information required by the European Union is provided in Table 4.20. The difficulty in obtaining this in-depth data in each ecosystem in which the organism is to be used can be readily imagined.

# Ethics

Genetic engineering stirs the primal emotions of right and wrong and fear of the unknown as discussed in the first part of this chapter. Yet, even if the moral objections to "playing the role of

## TABLE 4.20
### Regulatory Information for Environmental Release of Genetically Modified Organisms (European Union)

The organism:
  Organism modification methods
  Construction and implementation of the genetic change (insert or deletion)
  Purity of inserted information and other sequences/functions present
  Similarity in sequence, function, or location of the modified nucleic acid sequences to any known harmful
    sequences

Environmental impact:
  Potential for excessive population increase of organism in environment
  Competitive advantages of modified over unmodified organism in environment
  Anticipated mechanisms and consequences of interactions between modified and unmodified organisms
  Identification and description of non-similar organisms that might be effected by genetically modified organism
  Probability of shifts in biological interactions such as host shift after release
  Known or predicted effects on other organisms in the environment such as population changes
  Known or predicted involvement in biogeochemical processes
  Other potentially significant interactions with the environment

EEC Directives:
  90/219/EEC  Containment
  90/220/EEC  Release and Marketing
  90/679/EEC  Health and Safety of Workers Using Genetically Modified Organisms

Adapted from European Union directive 90/220/EEC. Council of European Communities, *Official Journal* L117/115: 15–27 (1990). In Kappeli, O., and Auberson, L., *Trends in Biotechnology* 15: 342–349 (1997).

God" are satisfied, there remain many questions about the social and personal impact of the knowledge made available by the technology. The following discussion is predicated on general notions of the mores, culture, politics, and economic environment prevalent in the United States. Priorities, accessibility, and the impact of genetic engineering technology are likely to vary in different societies as will the responses of their people. People's individual feelings of right and wrong may be in distinction to their more homogeneous responses on issues of wider scope such as concern over potential ecological disturbances. Is society ready for the disclosure of hidden susceptibilities and the insecurity of knowing but not truly understanding? The present information glut in areas of much less importance is already hard to handle for many people. Can the right of individuals to privacy of their medical information be protected and how much should be shielded? Can society resist the temptation to interpret probabilities as certainties, denigrating people whose genetic heritage condemns them to an uncertain but threatened future?

People's response to presently available biochemical and genetic tests suggests that just because the information is obtainable, it doesn't follow that it will or should be used. Simply banning genetic testing is similarly not a workable option. An unfortunate adjunct of our increasingly litigious society and the availability of genetic testing brings the potential for "wrongful birth" lawsuits by parents and "wrongful life" lawsuits brought by, or on the behalf of, an afflicted child against health care providers, if testing was not offered or appropriate counseling not provided. Leonard Fleck, professor of philosophy and medical ethics at Michigan State University's Center for Ethics and Humanities in the Life Sciences, aptly sums up the dilemma (Holoweiko, 1997): "We used to live in the age of genetic innocence, with no control over our genetic fate or that of our children. Now we live in the age of genetic responsibility."

## Genetic Testing

Approximately 3% of all children are born with a severe disorder generally considered to be genetic in origin. Most genetic diseases manifest early in life, particularly during prenatal development, although there are also a significant number of adult onset genetic diseases, such as the neurologic disease Huntington's chorea, or Huntington disease (HD). Some examples of prenatally diagnosable genetic diseases are given in Table 4.21.

Guidelines established by the Institute of Medicine of the National Academy of Sciences for the use of genetic testing depend upon the age of the individual being tested, with increased stringency mandated for those considered unable to make informed decisions. The criteria for allowing genetic screening for newborns are that there must be a clear benefit to the newborn, confirmatory tests must be available, and both treatment and followup must be available for affected individuals. This specifically forbids determining whether a newborn carries a predisposition to a disease as opposed to an actual genetic trait or disease. If a parent is a possible carrier of a disease gene, then testing of the parents is recommended.

In general, testing of children is indicated only if there is a curative or preventative treatment that must be applied early in life to be effective. Carrier status testing or testing for incurable or late-onset diseases (unless preventable by early treatment) is to be deferred to adulthood when the individual is presumed capable of informed decisions. The reasoning includes the concept

**Table 4.21**
**Examples of Prenatally Diagnosable Genetic Diseases**

| Disease | Incidence | Symptoms |
|---|---|---|
| Phenylketonuria | 1:10,000 | Mental retardation |
| Hemoglobinopathies | | Anemia |
|   Sickle cell | Common in African & Mediterranean | |
|   ß-thalassemia | Common in Mediterranean & Asia | |
| α1-antirypsin deficiency | 1:8,000 | Emphysema (lung), liver disease |
| Hemophilia A | 1:10,000 males | Bleeding disorders |
| Tay-Sach's Disease | 1:300,000 general population | Mental retardation |
| | 1:3,000 Ashkenazi Jews | |
| Gaucher's Disease | Rare in general population | Anemia, enlarged spleen |
| | 1:600 Ashkenazi Jews | |
| Glucose-6-phosphate | Variable, many mutations | Anemia |
| Dehydrogenase deficiency | | |
| Type II hyperlipidemia | 1:2,500 | Atherosclerosis, coronary heart disease |
| Familial hypercholesterolemia | 1:500 | Atherosclerosis, coronary heart disease |
| Duchenne muscular dystrophy | 1:3,000 | Muscle wasting |
| Cystic fibrosis | 1:2–3,000 | Lung failure, digestive malabsorption |
| Neurofibromatosis | 1:3,500 | Nervous system tumors |
| Adult polycystic disease | 1:5,000 | Kidney failure |
| Huntington's Disease | 1:10–12,000 | Uncontrolled movement-nervous system degeneration |
| Spinocerebellar ataxias | 1:25–50,000 | Movement disorders-nervous system degeneration |

that unrestrained childhood testing infringes on that individual's confidentiality rights (normally provided to adults). Knowledge of a child's genetic status risks stigmatizing his/her upbringing and relationships to family members, and raises life and health insurance issues. Testing children also intrudes on their future right to make their own decisions for testing as adults. Depending on the disease, eligible adults may, in fact, choose not to be tested. One-third of at-risk Huntington's disease individuals did not plan to make use of genetic testing (Adam et al., 1993). A Belgian study showed that while 75% of partners of at-risk individuals were in favor of predictive genetic testing, only 29% of the at-risk individuals were supportive. Thirty percent of pregnant at-risk individuals requested the testing on themselves, and only 18% actually had it performed (Evers-Kiebooms, 1990). Nancy Wexler, a Ph.D. clinical psychologist at the California-based Hereditary Disease Foundation, herself at risk for HD, remarked in 1984, "It's not a good test if you can't offer a treatment" (Lyon and Gorner, 1995). Eighteen percent of parents with a child afflicted with phenylketonuria were willing to undergo prenatal diagnosis in subsequent pregnancies (Barwell and Pollitt, 1987).

Eighty-one percent of women at increased risk for fragile X syndrome (Moorish and Abuelo, 1988), and eighty-two percent of carriers of hemophilia (Lajos and Czeizel, 1987) would choose prenatal testing under the same circumstances.

Arguments in support of genetic testing of children can be made on the basis of a parent's right-to-know and resolution of parental uncertainty as well as provision of lead time for psychological adjustment. There are also families in which siblings have already have been tested. Proponents of childhood testing also point out that based on genetic statistics, between 33% and 50% of those tested will be reassured by the results that they don't have the potential for the disease.

Newborn testing for catastrophic fetal disorders by looking for biochemical expression of the disease has been mandated by many states for decades (Table 4.22).

In contrast to genetic analysis, there has been little criticism of diagnostic biochemical type of testing of newborns. Biochemical tests more closely measure the clinical expression of the disease rather than simply a risk or probability.

Advances in genetic technologies made possible by recombinant DNA techniques are now making it possible to detect many more genetic abnormalities with higher sensitivity than with the biochemical tests. Table 4.23 provides a list of some of the techniques now in use along with their pros, cons, and estimated costs in full production.

**TABLE 4.22**
**U.S. Screening of Newborns for Genetic Disorders (1990)**

| Disorder | Number of states promoting screening[a] |
|---|---|
| Phenylketonuria | 52 |
| Congenital hypothyroidism | 52 |
| Hemoglobinopathy | 42 |
| Galactosemia | 38 |
| Maple syrup urine disorder | 22 |
| Homocysteinuria | 21 |
| Biotinidase deficiency | 14 |
| Adrenal hyperplasia | 8 |
| Tyrosinemia | 5 |
| Cystic fibrosis | 3 |
| Toxoplasmosis | 3 |

[a] Including District of Columbia, Puerto Rico, and U.S. Virgin Islands
Source: Council of Regional Networks for Genetic Services, 1992. In Andrews, L.B. et al., 1994, Assessing Genetic Risks, National Academy of Sciences, 68–69.

**Table 4.23**
**Genetic Testing Techniques—Scanning for Mutations**

| Technique | Sensitivity | #Samples/wk | Cost/sample | Pros | Cons |
|---|---|---|---|---|---|
| Nucleotide sequencing | 99% | 10 | $100 | Highly accurate | Expensive, labor-intensive |
| Chip technology | – | High | – | Potentially highly accurate and fast | Unknown general appl. |
| Protein truncation | 50–95% | 40–100 | $30–70 | Rapid, inexpensive | Variable accuracy, RNA-based |
| Single strand | 60–95% | ~8 | $200–500 | Easily performed | Variable accuracy, multiple conditions req'd |
| Mismatch cleavage detection | ~99% | 12 | $150 | High accuracy | Difficult to perform |
| Denaturing gradient gel electrophoresis | 99% | 40 | $100 | High accuracy, easy, inexpensive | Long optimization |

– = unknown
Source: *Nature Biotechnology* 15: 424–426, 1997, Eng, C. and Vijg, J.

So why isn't all testing done by genetic means? The reason is that the number of factors involved in producing most disease states makes it difficult to predict exactly who will develop the disease and when. The tests determine only the average probability of developing the disease at some time. Biochemical tests based on the abnormality only identify those that have clinical expression of the condition at the time of testing. They cannot predict those that will develop the disease later in life. One in twelve African Americans are carriers of the sickle cell trait, a hemoglobin variant that affects the red cell response to oxygen. If both parents carry that recessive trait, their child has a one in four chance of having sickle cell anemia. Yet, with prophylactic penicillin treatment to prevent infections the observed incidence of sickle cell anemia is actually one in six hundred. Thus, having the gene variant does not guarantee having the disease. Other tests for later onset genetic diseases are performed on adults suspected to be at risk because of a family history of a disorder after they have decided that they want to know whether they are affected or a carrier (Table 4.24).

Even though individual identified genetic diseases are quite rare (see Table 4.21 for some incidence figures), a sizeable number of people are involved in decisions about genetic testing. Thirty-seven percent of respondents to a survey by the National Opinion Research Center in 1990 (Roper Center for Public Research,

**Table 4.24**
**Examples of Late-Onset Genetic Disorders**

| Disease | Incidence | Pathology |
|---|---|---|
| **Monogenic** | | |
| Huntington's Chorea | 1:100,000 | Uncontrolled movement |
| Hemochromatosis | 1:500 (Caucasian) | Iron storage |
| Familial Hyper-Cholesterolemia | 1:500 | Atherosclerosis |
| Polycystic kidney disease | 1:500–1:5,000 | Renal failure |
| Inherited susceptibility to cancers | 1:200 | p53, breast cancers |
| **Multigenic** | | |
| Congestive heart disease | | Several major susceptibility genes |
| Hypertension | | Several genes regulating sensitivity |
| Cancers of complex origin | | Not inherited-somatic changes in tissue |
| Diabetes | | Numerous genes |
| Rheumatoid arthritis | | HLA antigens and others |
| Psychiatric disorders | | Schizophrenia, manic depressive disorder, panic disorders, Tourette's Syndrome, some alcoholism |
| Schizophrenia | | Genes on several chromosomes |
| Bipolar disorder | | Genes on several chromosomes |

1990) reported having an immediate family member who had, or was at risk for, or was a carrier of, a genetic disease. This surprising percentage is likely to be an overestimate because in identifying the disorder most people in the survey focused on physical deformities or on mental retardation that in many cases actually may have been environmental or traumatic and not genetic in origin. Just as for surveys reporting a high level of scientific illiteracy, 85% of the respondents to the above survey claimed to have read or heard little or nothing about genetic screening. Nevertheless, they expressed strong opinions about screening and were wary of the potential for misuse. A March of Dimes survey also found that 68% of respondents knew little or nothing about genetic testing and that 87% were similarly ill-informed about gene therapy (Louis Harris and Associates, 1992).

# Employment and Insurance

Employers in increasing numbers are using genetic testing in support of hiring decisions as an adjunct to a battery of psychological and standardized tests. The objective is to reduce the risk that a new employee will be transient or frequently absent, be disruptive, or increase the cost of workers' compensation coverage or medical insurance. Another use for such information would be identifying people with a propensity to allergies or a

genetic susceptibility to cancer and excluding them from working in environments with exposure to agents at levels that have
little effect on the general population. Opponents of this type of
testing demand that the employer should be responsible for reducing the environmental risk to permit even highly sensitive individuals to work there. Such uses of genetic testing argue that a
genetic predisposition can be equated with a preexisting condition. The logic of this reasoning is flawed since for most genetic
disorders the genetic marker reflects only a probability that a person will develop a disease, not that they have the disease. Opposition to the use of genetic testing for setting selection criteria that
are beyond an individual's control, such as race or genetic background, comes from the perception of many of the public that
such practices are unfair and discriminatory.

Fewer than half of the states have addressed the employment
issue, and the Federal government has only recently begun to consider legislation on the privacy of genetic information. As of March
1998, fourteen states (Arizona, Florida, Illinois, Iowa, Louisiana,
New Hampshire, New Jersey, New York, North Carolina, Oklahoma, Oregon, Rhode Island, Texas, and Wisconsin) restrict employer access to or use of genetic information. Some states are
more comprehensive than others in their protection. Florida and
Louisiana cover only sickle cell disease, and Oklahoma has only
set up a task force to make nonbinding recommendations to the
state legislature. The other eleven states with legislation forbid dis-
charging or refusing to hire an individual on the basis of genetic information. Seven states (Arizona, Iowa, New York, Oregon, Rhode
Island, Texas, and Wisconsin) forbid employer access to results of
genetic testing without an individual's consent, and some allow no
access at all. Iowa, New Hampshire, New Jersey, New York, Rhode
Island, Oregon, and Wisconsin forbid employers from requiring
genetic testing as a condition of hiring or continued employment.
New Hampshire, New Jersey, New York, and Wisconsin permit
susceptibility testing for occupation-related disorders if requested
by an individual seeking employment who might be worried
about his/her sensitivity to environmental conditions.

People are overwhelmingly opposed to using genetic testing
in employment decisions: 85% (National Opinion Research Center, 1990) and 80% (ABC News Poll, 1990). Seventy-five percent
supported the principle that tested individuals should retain sole
control over access to their genetic testing information (National
Opinion Research Center, 1990). The sentiment is mirrored in
Germany, where a survey showed that 75% were opposed to the

use of genetic testing in employment screening, with 23% favoring prohibiting genetic testing altogether (Hennen, Petermann, and Schmitt, 1993). However, 26% of the respondents in the German survey thought that people had a duty to have their genetic makeup tested.

A similar campaign is being waged in the U.S., to prevent or restrict the use of genetic information by the health and life insurance industries in determining the cost of covering individuals. In actuarial analysis, risk of a particular outcome is calculated for different categories of people that have been grouped by similarities based on information available to the company—age, high blood pressure, cholesterol, family history, smoking history, and other factors. The companies would like to use genetic testing to further stratify their clients to improve their ability to predict risk (and, some people believe, simply to exclude people with the potential to develop expensive diseases). While this would protect the company against losing money on payments for exceptionally high risk individuals, charging premiums adjusted for the individual risk would make it even more difficult for people with chronic diseases to find and retain insurance coverage at an affordable cost. Denial of insurance coverage is not an exceptional event. HMOs in 1987 denied membership to 24% of individual applicants (not group applicants, who are usually covered through their place of employment or other organization).

Federal government legislation passed in 1996 exists to bar insurers from using genetic predisposition information to deny coverage or charge high rates to individuals based on a "preexisting condition" unless the disease is already clinically active. This legislation protects some 150 million Americans insured through group plans primarily at their places of employment. Thirteen million Americans with individual health care insurance policies enjoy no such refuge. Fear of having their genetic information used against them looms large for many people. An NIH study found that 32% of women approached to take part in a genetic breast cancer detection trial did not participate, mostly due to fear of discrimination and lost privacy. One woman, whose family includes nine breast cancer cases in three generations, is waiting until legislation guaranteeing nondiscrimination is on the books. "My doctor said to me, 'If you get this test done, your daughter is not going to get insurance'" (Ann Arbor News, 1998). In the absence of an effective state or national health care program, how can the needs of people to have affordable health care regardless of their genetic and economic status be provided for? Assuming that this is approached through insurance plans and that the companies are entitled to use all available information to match each insuree's premium payment to his/her individual risk, there are several possibilities. One compromise would be for all of the insured to pay more to subsidize those more likely to require benefits. This would yield essentially the present situation when specific genetic information is not known. Another would be to provide an adequate basic level of coverage for everyone without the use of genetic information, with the opportunity for additional coverage using all available information including genetic testing to set the rates. Still another option was suggested by Alexander Tabarrok, assistant professor of economics at Ball State University (Brostoff, 1996). Before opting for genetic testing, people could purchase insurance against finding that they carry a disease-causing gene. This insurance, if a person tested positive for a genetic disease marker, would cover the higher premiums charged by an insurance company and/or the higher health care costs.

The rapidly expanding capabilities and use of genetic testing in personal health and employment decisions has prompted the drafting of legislation to control abuses of the new information. A series of bills (House of Representatives H.R. 306, H.R. 341, H.R. 1815; Senate S. 89 and S. 422), designed to control the circumstances under which DNA samples may be collected, analyzed, and the information disclosed, were introduced in the 105th Congress. They specified prohibition of discrimination for employment or health insurance on the basis of genetic information. This legislation will form the backbone of individual and family member privacy law. Some people feel that blocking access to genetic information will be futile and that regulation would be more effective if targeted against the use of the information. Hastily proposed in early 1997, none of this legislation has been enacted through mid-1998 due to concern over the potential limitation on access that could include genetic information such as that collected under informed consent in clinical trials. Industrial organizations such as the Pharmaceutical Manufacturers Association feel that such blanket denial of access will severely restrict the use of anonymous genetic information to devise new medical therapies.

## Proposed Legislation

S. 422, the Genetic Confidentiality and Nondiscrimination Act of 1997, defines the circumstances under which DNA samples and genetic information may be collected, stored, and analyzed, and the rights of individuals and persons with respect to genetic information, as well as providing protection from genetic discrimination. It was introduced by Mr. Domenici, Mr. Jeffords, and Mr. Dodd on March 11, 1997, and was referred to the Committee on Labor and Human Resources.

H.R. 306, the Genetic Information Nondiscrimination in Health Insurance Act of 1997, amends the appropriate laws to prevent discrimination in employment or in health insurance against individuals and their family members on the basis of genetic information or on a request for genetic services. It was introduced by Ms. Slaughter and 43 other representatives on January 7, 1997, and was referred to the Committees on Commerce, Ways and Means, and Education and the Workforce.

S. 89, the Genetic Information Nondiscrimination in Health Insurance Act of 1997, is a companion bill in the Senate to H.R. 306. Introduced by Ms. Snowe on January 21, 1997, it was referred to the Committee on Labor and Human Resources.

H.R. 341, the Genetic Privacy and Nondiscrimination Act of 1997, establishes limits on the disclosure and use of genetic information including prohibiting discrimination in employment or by health care insurers. Introduced by Mr. Stearns and 9 other representatives on January 7, 1997, this bill was referred to the Committees on Commerce, Government Reform and Oversight, and Education and the Workforce.

H.R. 1815, the Medical Privacy in the Age of New Technologies Act of 1997, provides for the protection of private medical information including genetic data as well as any produced in the future by new technologies. It codifies establishment of safeguards and rules for disclosure, particularly addressing the increased accessibility of medical information through computerized networks. Introduced by Mr. McDermott and 9 other representatives on June 5, 1997, this bill was referred to the Committee on Commerce and the Committee on Government Reform and Oversight.

A twenty-first century concern is already here, the direct marketing of genetic testing to the public. Some commercial laboratories such as Myriad Genetic Laboratories, Inc. (Salt Lake City, Utah) and Oncormed, Inc. (Gaithersburg, Maryland), that offer breast cancer gene screening, do so only to a group of select clients. They also provide instruction to physicians about how to screen prospective patients and provide educational materials for medical professionals and consumers. In addition, these companies assist physicians in locating genetic counseling for their high risk patients. On the other hand, there remains the potential for marketing to the general public without providing adequate interpretation and counseling to make use of the results, a situation characterized by Glenn McGee (Center for Bioethics, University of Pennsylvania ) as "drive-through genetic testing." As part of the Human Genome Project, the ELSI Task Force on Genetic Testing issued a series of recommendations in May, 1997 (Lewis, 1997). These included:

1) Institutional review board approval of predictive genetic tests;
2) Outside review of clinical utility and validation data provided by test developers;
3) Enhanced genetic training of health care professionals including nurses, social workers, and public health workers whose patients have diseases with substantial inherited components;
4) Development of stringent criteria for predictive genetic tests;
5) Requiring hospitals or managed care organizations to show evidence of competency to interpret genetic tests and provide counseling before ordering genetic tests;
6) Improved transmission of new information on rare genetic diseases to physicians.

Genetic testing may eventually be found on supermarket shelves alongside the blood glucose and pregnancy home testing kits. Gamera Bioscience Corp. (Medford, Massachusetts) is developing the "LabCD System," a compact disk player–sized unit that will perform several different genetic tests simultaneously. While initially targeting test manufacturers, the company has an eye on physicians' offices and even the home, Gamera chief executive officer Alec Mian has indicated (Lewis 1997). How will enforcement of the laudable safeguards of the ELSI Task Force on Genetic Testing against misuse and abuse of genetic information be assured?

## DNA Forensics

Although the FBI maintains forensic DNA capability, several commercial companies have specialized in performing forensic DNA analysis that is beyond the scope of most forensic laboratories. Cellmark Diagnostics, a subsidiary of Imperial Chemical

Industries, which opened labs in Germantown, Maryland, in 1987, pioneered the "DNA fingerprinting" technology based on RFLP analysis. Lifecodes Corporation, Valhalla, New York, also used an RFLP technology while Forensic Science Associates, Richmond, California, has pursued polymerase chain reaction (PCR) amplification and marker alleles of known genes such as human leukocyte antigen (HLA) DQ-1 locus. Gennan Corporation of Akron, Ohio, and Genescreen of Dallas, Texas, have also been identified as commercial forensic DNA analysis laboratories by the U.S. Office of Technology Assessment. Through January, 1990, 185 court cases reported using DNA typing evidence in the prosecution or defense (Table 4.25). Often these cases were crimes of violence where there was no living witness, thus rendering the DNA evidence especially crucial in obtaining a conviction.

The tendency to collect large databases of information operates in law enforcement perhaps with more energy than in any area other than the military. Law enforcement officials cite the high rate of recidivism among convicted violent offenders as justification for the maintenance of databases to aid in identification and swift prosecution. In 1989 Bureau of Justice statistics reported that 62.5% of prisoners (59.6% of the violent criminals) released in 1983 had been rearrested within 3 years, with 41.4% being returned to prison. Among the violent offenders, rapists were 10.5 times more likely than other released offenders to be rearrested for rape; murderers about 5 times more likely than other offenders to be rearrested for homicide. While these statistics would argue powerfully for databases to protect society, it is noteworthy that in the same study only 6.6% of the released rapists were rearrested for rape and 7.7% of the killers were rearrested for murder (Bureau of Justice, 1989). The civil rights of convicted criminals, particularly felons, are legally curtailed compared with those of the average citizen. The Privacy Act of 1974 (5 U.S.C. 552a) governing data collection and access to information about most people in federal databases specifically exempts criminal justice agency record systems from many provisions (5 U.S.C. 552a(b)(7), (c)(3), and (j)(2)). States control the privacy of nonfederal criminal history databases, ranging from complete public accessibility (Florida) to sealed records (Massachusetts). Pressed by rising crime statistics, legislatures in a number of states (as of 1990) passed legislation to require or propose the collection of DNA samples and test results from certain convicted offenders, particularly crimes of a sexual nature. The creation of a national DNA database could aid the criminal

## TABLE 4.25
### Reported Uses of DNA Typing and DNA Databank Legislation (1990)

| State | Number of cases | (P)roposed or (R)equired DNA databanking from certain convicted offenders |
|---|---|---|
| Alabama | 6 | – |
| Alaska | b | – |
| Arizona | 3 | – |
| Arkansas | 1 | – |
| California | 8 | R |
| Colorado | 7 | R |
| Connecticut | 1 | P |
| Delaware | 1 | – |
| District of Columbia | b | – |
| Florida | 25 | R |
| Georgia | 4 | – |
| Hawaii | 1 | – |
| Idaho | 1 | – |
| Illinois | 1 | P |
| Indiana | 3 | R |
| Iowa | 2 | R |
| Kansas | 5 | – |
| Kentucky | c | – |
| Louisiana | 1 | – |
| Maine | b | – |
| Maryland | 11 | – |
| Massachusetts | 1 | P |
| Michigan | 5 | P |
| Minnesota | 2 | – |
| Mississippi | 4 | – |
| Missouri | 2 | – |
| Montana | 1 | – |
| Nebraska | c | – |
| Nevada | c | R |
| New Hampshire | 2 | – |
| New Jersey | b | – |
| New Mexico | b | – |
| New York | 17 | – |
| North Carolina | 4 | – |
| North Dakota | c | – |
| Ohio | 10 | P |
| Oklahoma | 3 | – |
| Oregon | 1 | – |
| Pennsylvania | 9 | – |
| Rhode Island | b | – |
| South Carolina | 4 | – |
| South Dakota | 1 | R |
| Tennessee | 1 | – |
| Texas | 18 | – |

**TABLE 4.25** *continued*

| State | Number of cases | (P)roposed or (R)equired DNA databanking from certain convicted offenders |
|---|---|---|
| Utah | 1 | – |
| Vermont | b | – |
| Virginia | 10 | R |
| Washington | 4 | R |
| West Virginia | 1 | – |
| Wisconsin | 1 | R |
| Wyoming | c | R |

[a] Cases in which DNA evidence was admitted by a court or used to obtain a plea prior to an admissability hearing. Two military cases are not reported in this table.
[b] Cases pending, DNA used to determine innocence only, DNA evidence withdrawn, or DNA evidence used in investigation but no prosecution.
[c] None identified.
Source: Office of Technology Assessment, 1990. *Genetic Witness.*

justice system and the military, but it would be essential in many people's minds to restrict the use of that information. By the year 2000 the U.S. military would create a huge DNA sample library and database to aid in identifying remains, but critics question how long the DNA will be held and under what conditions information from that DNA could be used.

# Human Cloning

On February 22, 1997, the successful differentiated cell nuclear transfer (or so-called cloning) of Dolly, a Finn Dorset lamb, was announced (Wilmut, et al., 1997). Shortly thereafter genetically nonidentical rhesus monkeys were cloned at the Oregon Regional Primate Research Center, using embryonic nuclear transfer to enucleated mature oocytes. This was followed by several other similar experiments with other animals, and elicited an immediate call for legislation to control, if not outright ban, the application of similar technology to humans. On February 27, 1997, a legislative bill S. 368, restricting the use of federal funding for research related to cloning, was introduced into the Senate. This was swiftly followed on March 5, 1997, by the introduction of H.R. 922 into the House of Representatives with a similar intent. Significantly, this was supplemented with H.R. 923, which would make it a civil crime for a person to conduct cloning of a human. As introduced, these bills contain little more than the initial statement of purpose.

The details will be added as the bills make their way through the legislative process. A note of caution about blanket bans on cloning research was injected by people trying to keep in sight the broader vision for society. The chairman of the genetics subcommittee of the National Bioethics Advisory Commission, Thomas H. Murray, Ph.D. (director of the Center for Biomedical Ethics, Case Western Reserve University School of Medicine, Cleveland, Ohio), remarked at the March 5 House Science Subcommittee meeting, "It is essential that we keep in view the possible scientific, and ultimately the human, benefits of research on animal cloning, benefits best described by other witnesses before you today. It is important that our public policy response to research on the cloning of animals not be swept along by our concern to prevent what we will judge to be the ethical dangers of human cloning."

H.R. 922, the Human Cloning Research Prohibition Act, prohibits the "expenditure of federal funds to conduct or support research on the cloning of humans." Introduced by Mr. Ehlers on March 5, 1997, it was referred to the Committee on Commerce and the Committee on Science.

H.R. 923, the Human Cloning Prohibition Act, makes it "unlawful for any person to use a human somatic cell for the process of producing a human clone." It imposes a civil penalty of up to $5,000. Introduced by Mr. Ehlers on March 5, 1997, it was referred to the Committee on Commerce.

S. 368, "To prohibit the use of Federal funds for human cloning research," defines cloning to be "the replication of a human individual by the taking of a cell with genetic material and the cultivation of the cell through the egg, embryo, fetal, and newborn stages into a new human individual." Introduced on February 27, 1997, by Mr. Bond and Mr. Ashcroft, it was referred to the Committee on Labor and Human Resources.

Before the passage of legislation, a voluntary moratorium on human cloning was generally agreed on by scientists, medical professionals, legislators, and industry. Though the moratorium lacks the force of law, the initial use of recombinant DNA technology in the early 1970s serves as historical precedent for the effectiveness of such voluntary restrictions. At that time, potentially risky experiments were delayed for several years until detailed guidelines could be developed and safety issues had been resolved. It is likely that analogous guidelines will evolve for cloning. News opinion polls conducted soon after the cloning announcements showed a significant amount of concern:

56% (America Online 1998) and 93% (Times-CNN 1999) of Americans opposed human cloning. At the request of President Clinton, the National Bioethics Advisory Commission called for immediate action on:

1) "A continuation of the current moratorium on the use of federal funding in support of any attempt to create a child by somatic cell nuclear transfer.
2) An immediate request to all firms, clinicians, investigators, and professional societies in the private and nonfederally funded sectors to comply voluntarily with the intent of the federal moratorium. Professional and scientific societies should make clear that any attempt to create a child by somatic cell nuclear transfer and implantation into a woman's body would be at this time an irresponsible, unethical, and unprofessional act."

In order to address the general public concern about human cloning, scientific societies such as the Federation of American Societies for Experimental Biology (FASEB) adopted specific resolutions supporting the moratorium and called upon scientists to provide input to ensure that imprecise or misused technical language in antihuman cloning legislation did not compromise vital biomedical research. For instance, "cloning" is a general term that includes general recombinant DNA handling as well as somatic cell nuclear transfer. Research involving "cloning" the human insulin receptor and expressing it in a cultured cell line to learn more about how insulin works and what might be going wrong in diabetes could be prohibited by sloppily written legislation. The FASEB Public Affairs Executive Committee adopted the following resolution on September 10, 1997:

Resolved: The Federation of American Societies for Experimental Biology (FASEB) adopts a voluntary five-year moratorium on cloning human beings, where "cloning human beings" is defined as the duplication of an existing or previously existing human being by transferring the nucleus of a differentiated, somatic cell into an enucleated human oocyte, and implanting the resulting product for intrauterine gestation and subsequent birth.

Although seemingly needlessly complex, the detail in the

legislation is essential; it avoids impacting cloning of human genes for medical research as well as the scenario in which a woman is prosecuted for "cloning" by bearing monozygotic (identical) twins. In fact, mammalian cloning has been widespread for a number of years with the cloning and implantation of prize-winning cattle embryos. It is not entirely clear to some people why cloning people presents such a problem since in effect, with identical twins, there are many "clones" already around. Society deals with them as individual persons with little remark. They are identical genetically but this, as has been pointed out in a previous section, addresses only potentiality, not actuality. A number of scientists and the general public, particularly in the United States, now wonder whether our leaders were stampeded into poorly thought out legislative action. The broadly written Human Cloning Prohibition Act, S. 1061, which would have made human somatic cell nuclear transfer a criminal offense, punishable by fines and a 10-year prison sentence, had been hurriedly introduced. During the February 11, 1998, debate on S. 1061, Senator Connie Mack—himself stricken with cancer, with a wife and daughter similarly diagnosed, and whose parents and brother died of cancer—offered an impassioned plea:

> I appeal to you, don't get drawn into this debate that we should pass this legislation because we want to stand up and make a statement that we are against cloning. We are all against human cloning. What I am asking you to do is to vote no on cloture so we will have an opportunity to hear from those patient groups that want to represent people like myself, represent families that have been affected like my family has been affected....

The cloture vote was to cut off debate, forcing an immediate vote on the bill without the discussion of a full hearing that would present the information needed to make an informed decision on this technical yet highly emotional issue. Senator Robert C. Byrd (D-W.Va.) offered the caution:

> Who can say with any comfort what the impact may be on important research aimed at dread diseases? Doesn't important and potentially far-reaching legislation such as this at least warrant hearings before we proceed? This legislation could have unintended and detrimental consequences.

By a count of 42 to 54, short of the 60 votes needed to end de-bate, the Senate in effect decided to continue the discussion. A substitute Human Cloning Prohibition Bill, S. 1062, jointly spon-sored by Senators Diane Feinstein (D-California) and Edward Kennedy (D-Massachusetts), that bans the implantation of the product of nuclear transplantation in a woman's womb but not the nuclear transplantation technology itself, will soon be con-sidered. While being explicit about restricting human somatic cell nuclear transplantation for the purpose of creating a child, specific protections for nuclear transfer in nonhuman animals, general DNA cloning, and vital medical research are provided.

BIO, the Biotechnology Industry Organization, representing more than 745 biotechnology companies, academic institutions, and state biotechnology centers in 46 states and more than 25 countries, agrees that cloning of a human being is "plainly inap-propriate." Carl B. Feldbaum, president of BIO, issued the fol-lowing statement (in part):

> BIO continues to support President Clinton's and the National Bioethics Advisory Commission's (NBAC) moratorium on the cloning of a human being. Cloning a human being poses major ethical and moral ques-tions, as well as deeply troubling medical safety is-sues... the public has come to learn of the benefits of cloning cells, genes, and tissues, techniques that have been ongoing for 20 years. While these techniques do not lead to the cloning of a human being, they do have promise, for example, to enable us to regenerate spinal cord tissue for accident victims and skin for burn victims. It is essential that any restrictive legisla-tion that may be adopted by Congress or state legisla-tures on the cloning of a human being, recognize the biomedical benefits of these existing cloning proce-dures and protect and encourage them (Feldbaum, 1998).

The international reaction to Dolly was swift and severe, similar to that in the United States. Great Britain banned human cloning outright, while a number of European countries, includ-ing Germany and Italy, either called for worldwide bans or insti-tuted temporary moratoria on human cloning experiments.

# Summary

Genetic engineering is poised to make its greatest impact over the next few decades. It is hard to judge the balance between the relative good and bad potential of the technology. To this point we have been able to catch only a glimmer of the potential advances to be made with this newfound power to see into and to influence Nature on a scale never before imagined. This power is frightening to many people. It has forced us to think both about our place in the world ecological web and about our relationships to one another. Powerful tools to improve our understanding of the chemistry of life and to probe what goes wrong in disease are the major accomplishments of the first 25 years of genetic engineering. New medicines from this, certainly, or even eradication of a genetic disease, still in the future, are positive contributions. Less understood and greatly feared by many are potential repercussions in the environment from the wide-scale release of genetically modified plants, animals, and microbes in industrial, agricultural, or environmental remediation applications of genetic engineering. Acceptable control measures remain to be developed. Negative repercussions of genetic engineering, at least until society figures out how it wants to regulate it, are most notable in the genetic testing and forensic realms. Controlling the application of the information obtained from the testing will determine whether the impact of genetic engineering will be good or bad.

Genetic engineering technology challenges some of our deepest feelings about our private selves and makes us reevaluate our moral tenets and religious beliefs. It gives us access to more information about ourselves than we really want to know or, especially, to share with others. Humans demonstrate a need for some mystery in their lives and will invent the requisite amount when necessary. Knowing your genetic potential, though it reflects only a possibility, is unsettling, especially if people start to act on the basis of that information. There seems for most people to be a comfort zone in which some things must be left to chance to give a person the room to control his or her own destiny. The biggest and most far-reaching impact of genetic engineering will be social: how to guarantee the privacy of the individual. Living in the age of genetic responsibility will not be easy. There is too much information, too many choices, too many decisions to be made.

## References

ABC News. 1990. Genetic Engineering Poll (April). Radnor, PA: Chilton Research Services.

Adam, S., et al. 1993. "Five Year Study of Prenatal Testing for Huntington's Disease: Demand, Attitudes, and Psychological Assessment." *Journal of Medical Genetics* 30: 549–556.

Aldridge, S. 1997. "Researchers Tackling Bioremediation Challenge in Eastern Europe." *Genetic Engineering News* (October 1): 1.

American Association for the Advancement of Science. 1998. Philadelphia: Annual Meeting (February).

Barwell, B.E., and R. J. Pollit. 1987. "Attitudes des parents vis-à-vis du diagnostic prenatal de la phenylcetonurie." *Archives of French Pediatrics* 44: 665–666.

Biotechnology and the European Public Concerted Action Group (BEPCAG). 1997. "Europe Ambivalent on Biotechnology." *Nature* 387: 845–847.

Brostoff, S. 1996. "Genetic Testing Prompts New Insurance Coverage Proposal." *National Underwriter Property and Casualty-Risk and Benefits Management*, 16 (April 15): 33.

Bureau of Justice. 1989. *A Survey of Recidivism among Prisoners Released in 1983*. Washington, DC: U.S. Department of Justice.

Dibner, M. D. 1998. "Biotech Job Market Changing." *Genetic Engineering News* 1 (January): 12, 34.

Eisiedal, E. F. 1997. *Biotechnology and the Canadian Public: Report on a 1997 Survey and Some International Comparisons*. Calgary, Alberta, Canada: University of Calgary.

Evers-Kiebooms, G. 1990. "Predictive Testing for Huntington's Disease in Belgium." *Journal of Psychosomatic Obstetrics and Gynecology* 11: 61–72.

Feldbaum, Carl B. 1998. From *PR Newswire*, website http://www.prnewswire.com (January 7).

Fox, J. L. 1996. "BASIC Takes Bioremediation Public." *Nature Biotechnology* 14: 1077.

"Genetic Discrimination Laws Faltering." 1998. *Ann Arbor News* (April 11).

Glass, D. J. 1997. "Evaluating Phytoremediation's Potential Share of the Site-Remediation Market. *Genetic Engineering News* 17(17): 43.

Harris, Louis, and Associates. 1992. *Genetic Testing and Gene Therapy: National Survey Findings*. White Plains, NY: March of Dimes.

Hennen, L., T. Petermann, and J. Schmitt, 1993. "TA-Projekt 'Genomanalyse'—Chancer und Risiken Genetische Diagnostik." *TAB-Arbeitsbericht* 18.

Hoban, T. J., and P. A. Kendall. 1992. *Consumer Attitudes about the Use of Biotechnology in Agriculture and Food Production.* Raleigh: North Carolina State University.

Holoweiko, M. 1997. "Telling Right from Wrong Is Going to Get a Lot Harder." *Medical Economics* 74(14): 101.

Lajos, I., and A. Czeizel. 1987. "Letter to the Editor: Reproductive Choices in Hemophiliac Men and Carriers." *American Journal of Medical Genetics* 28: 519–520.

Lee, T. R., C. Cody, and Plastow. 1985. *Consumer Attitudes towards Technological Innovations in Food Policy.* Guildford, England: University of Surrey.

Lewis, R. 1997. "Genetic Testing for Cancer Presents Complex Challenges." *The Scientist* 11(20).

Lyon, J., and P. Gorner. 1995. *Altered Fates.* New York: W. W. Norton.

Macer, D. 1998. *Attitudes to Biotechnology in New Zealand and Japan in 1997 (Eurobarometer Survey).* Christchurch, New Zealand: Eubios Ethics Institute. http://www.biol.tsukuba.ac.jp/~macer/biotechnology.html

Moorish, D. L., and D. Abuelo. 1988. "Counseling Needs and Attitudes toward Prenatal Diagnosis and Abortion in Fragile X Families." *Clinical Genetics* 33: 349–355.

National Opinion Research Center. 1990. *U.S. General Social Survey.* Chicago.

Office of Technology Assessment. 1987. *New Developments in Biotechnology: Public Perceptions of Biotechnology* (OTA-BP-BA-45). Washington, DC: U.S. Government Printing Office.

Roper Center for Public Research. 1990. *Genes, Social Science Survey, 9/1990.* Storrs, CT: Public Opinion Online Database.

Thompson, P. B. 1997. "Food Biotechnology's Challenge to Cultural Integrity and Individual Consent." *Hastings Center Report* (27)4: 34–38.

Wilmut, I., et al. 1997. "Viable Offspring Derived from Fetal Adult Mammalian Cells." *Nature* 385: 810–830.

# Directory of Organizations and their Publications

5

This chapter describes U.S. and international organizations that deal with medical genetics, genetic engineering, or the impact of the technology on society. Brief background information on the organization, the purpose of the group, address, contact information, services provided, and publications of the organization are included. For the most part, these are organizations that are open to the public. Many organizations for professionals engaged in various aspects of genetic engineering have been omitted.

**Alliance of Genetic Support Groups**
4301 Connecticut Avenue NW, Suite 404
Washington, DC 20008
(202) 966-5557; (800) 336-GENE(4363)
Fax: (202) 966-8553
E-mail: info@geneticalliance.org
Website: http://www.geneticalliance.org
Contact: Executive director

The Alliance of Genetic Support Groups, founded in 1985, is composed of volunteer genetic organizations, professionals, and interested individuals who support individuals with genetic disorders and their families by facilitating interactions with government agencies, health care professionals, and service providers. It provides technical assistance and information for referrals to the support groups and the public. The Alliance maintains a database for its membership and offers a recognition award annually.

*Publications: Alliance Alert* (monthly newsletter), *Alliance Health Insurance Resource Guide, Directory of Voluntary Genetic Organizations and Related Resources, Informed Consent: Participation in Genetic Research Studies, Media Reporting in a Genetic Age, Alliance Resource Guide on Peer Support Training Programs.*

**American Genetic Association**
P.O. Box 257
Buckeystown, MD 21717-0257
(301) 695-9292
Fax: (301) 695-9292

A largely professional organization, the American Genetic Association includes biologists, zoologists, geneticists, botanists, and others exploring basic and applied research in genetics. Founded in 1903, it was formerly called the American Breeders Association.

*Publication: Journal of Heredity.*

**American Parkinson Disease Association**
1250 Mylan Blvd., Suite 48
Staten Island, NY 10305
(718) 981-8001; (800) 223-2132
Fax: (718) 981-4399
Contact: Vice president

Since 1961 this association has worked to alleviate the suffering of affected individuals and their families by subsidizing information and referral centers and providing funds for research into cures for the disease. Counseling services are offered to patients and families through 51 information and referral centers. It maintains a library, not open to the public, and offers various awards for contributions to patients' well-being and advancements in research on Parkinson Disease.

*Publications: American Parkinson Disease Association Newsletter* (quarterly), booklets on dealing with Parkinson Disease.

**American Society of Human Genetics**
9650 Rockville Pike
Bethesda, MD 20814-3998
(301) 571-1825

Fax: (301) 530-7079
Contact: Executive director

The American Society of Human Genetics, founded in 1948, is one of the oldest professional genetics organizations. Its members are involved in various aspects of medical genetics, from basic research to clinical applications.

*Publications: American Journal of Human Genetics, Guide to Human Genetics Training Programs in North America.*

**American Society of Plant Physiologists**
15501 Monona Drive
Rockville, MD 20855-2768
(301) 251-0560
Fax: (301) 279-2996
E-mail: skelly@aspp.org
Website: http://www.ophelia.com/Ophelia/pgr/index.html
Contact: Executive director

The American Society of Plant Physiologists is a professional organization of scientists interested in plant physiology. It is a source of plant expertise, including traditional biology and genetic engineering.

*Publication: Plant Physiology.*

**Applied Research Ethics National Association (ARENA)**
132 Boylston Street, 4th Floor
Boston, MA 02116
(617) 423-4112
Fax: (617) 423-1185
Contact: President

ARENA was founded in 1986. It consists of researchers, administrators, and professionals interested in bioethics, including members of institutional review boards, hospital ethics committees, patient advocacy groups, and institutional animal care and use groups, and primarily serves these groups. The group monitors federal legislation and public policy issues and provides expert consultation to members. It maintains a reference library and computerized mailing lists.

*Publications: ARENA Newsletter* (quarterly) and various brochures.

**Association of Birth Defect Children (ABDC)**
827 Irma Avenue

Orlando, FL 32803
(407) 245-7035; (800) 313-ABDC
Fax: (407) 245-7035
Contact: Executive director.

A multinational organization formed in 1980 of parents, health care professionals, other individuals, and organizations and services concerned with birth defects, particularly those associated with environmental exposure. ABDC conducts an educational campaign warning of the effects of prenatal exposure to environmental agents, sponsors studies and compiles statistics, and monitors and engages in legislative activities. The Association maintains a computer database, a National Birth Defect Registry, and parent matching to support services.

*Publications: ABDC* (quarterly newsletter), *Why My Child?* and videotapes.

**A-T Medical Research Foundation**
5241 Round Meadow Road
Hidden Hills, CA 91302
(818) 704-8146
Fax: (818) 704-8310
Contact: President

This multinational organization established in 1989 funds research on ataxia telangiectasia, a genetic neurodegenerative disease. The group maintains a reference library of clippings and articles. It also awards a grant to support research.

*Publication:* Newsletter.

**Behavior Genetics Association**
Division of Biostatistics, Box 8067
Washington University School of Medicine
660 S. Euclid
St. Louis, MO 63110
(314) 362-3642
Fax: (314) 362-2693
Contact: Secretary

A multinational organization of teachers and researchers studying the interrelationship of genetic mechanisms and animal and human behavior. Since 1971 it has encouraged and aided education and training of workers in the field and dissemination and interpretation of information on behavior and genetics to the public.

*Publications: Behavior Genetics* (bimonthly); also maintains *BGA Net,* an electronic bulletin board.

**Bioindustry Association**
1 Queen Anne's Gate
London SW1H 9BT
England
(71) 957-4600
Fax: (71) 957-4644
Contact: Executive director

A national organization of individuals interested in biotechnology development in the U.K. It seeks out contacts with trade associations and biotechnology organizations worldwide and represents members' interests in regulatory affairs, government policy, and financing. In support of its mission, the association conducts educational workshops, seminars, and conferences and maintains a reference library, a computer database, and e-mail facilities.

*Publications: UK Biotechnology* (biennial handbook) and *Biotechnologists' Book of Abbreviations and Acronyms.*

**Biotechnology Industry Organization (BIO)**
1625 K Street NW, Suite 1100
Washington, DC 20006
(202) 857-0244; (800) 255-3304
Fax: (202) 857-0237
Contact: President

BIO, founded in 1993, is an organization of industrial firms in the United States that are engaged in human health care, animal science, chemical production, and industrial fermentation processes. It also includes equipment manufacturers and servicers involved with recombinant DNA or hybridoma/immunologic technologies. Its mission is to provide information on biotechnology issues regarding various problems facing the industry such as U.S. and international regulations, patents, and financing opportunities. The group interacts with Congress, a variety of government and nongovernment agencies, regulatory bodies, and the public. BIO maintains a resource library and provides a recognition award.

*Publications: BIO Bulletin* (periodic), *BIO News* (bimonthly newsletter).

**Chromosome 18 Registry and Research Society**
6302 Fox Head
San Antonio, TX 78247
(210) 657-4968
Fax: (210) 657-4968
Contact: President

This multinational organization, founded in 1990, strives to locate individuals with chromosome 18 disorders and to educate the families and the public about prognosis and treatment. The group seeks to encourage research in the area and to link affected families and their physicians to the research community.

*Publication: Chromosome 18 Communique* (quarterly newsletter).

**Council for Responsible Genetics**
5 Upland Road, Suite 3
Cambridge, MA 02140
(617) 868-0870
Fax: (617) 491-5344
Contact: Executive director

This multinational watchdog group founded in 1983 monitors the biotechnology industry with concern for the social implications of genetic technology development. Primary issues of concern are genetically engineered foods, military uses of biotechnology, and issues of genetic discrimination. The Council maintains a reference library and a speakers bureau.

*Publication: GeneWatch* (bimonthly newsletter).

**Crop Science Society of America**
677 S. Segoe Road
Madison, WI 53711
(608) 273-8080
Fax: (608) 273-2021
E-mail: NRhodehamel@agronomy.org
Contact: Chairman

This is a professional society of individuals interested in issues surrounding plants and their cultivation as crops.

**Eugenics Special Interest Group**
P.O. Box 138
East Schodack, NY 12063
(518) 732-2390

Contact: Coordinator

The Eugenics Special Interest Group, founded in 1982, brings together people interested in issues of human genetics and population. The organization studies how children can be born healthy and live healthy lives.

*Publication: Eugenics Special Interest Group Bulletin* (periodic).

**European Federation of Biotechnology (EFB)**
c/o DECHEMA Postfach 15 01 04
60061 Frankfurt, Germany
(69) 756-4209
Fax: (69) 756-4201
Telex: 412490 DCHA D
Contact: General secretary

The EFB is a multinational organization representing technical and scientific organizations in 26 European and non-European countries whose purpose is to further biotechnology. Founded in 1978, it is concerned with animal and plant cell culture, production and processing of biomaterials, and environmental and safety issues.

*Publications: Annual Report of the EFB, EFB Newsletter* (English), *Proceedings of European Congress of Biotechnology* (periodic).

**Fanconi's Anemia Research Fund**
1902 Jefferson St., Suite 2
Eugene, OR 97405
(541) 687-4658; (800) 828-4891
Fax: (541) 687-0548
Contact: Family support contact

Since 1989 the Research Fund has provided support and networking advice to families, and updated them on new research findings. It operates a support group and conducts fundraisers for research. The organization maintains a library open to the public.

*Publications: FA Family Directory, FA Family Newsletter* (semiannual), *Fanconi Anemia: A Handbook for Families and their Physicians.*

**Foundation on Economic Trends**
1600 L Street NW
Washington, DC 20036-5603
(202) 466-2823
Contact: Director
E-mail: jrifkin@blarg.net

This foundation serves a watchdog function on technology and society issues in general and has entered on occasion into litigation to challenge government and industry policies. Founded by Jeremy Rifkin, a noted activist, it has supported a variety of his causes, including global economics, global warming, workers' rights, the impact of information technology, and genetic engineering of crops and food.

**Henry A. Wallace Institute for Alternative Agriculture**
9200 Edmonston Rd., Suite 117
Greenbelt, MD 20770
(301) 441-8777
Fax: (301) 220-0164
Contact: Executive director

Founded in 1983, this organization bands together farmers, agricultural scientists, educators, nutritionists, and consumers seeking to promote production methods that are economically viable, environmentally sound, and sustainable through development of educational programs and scientific research. It monitors and reports on related government activities.

*Publications: Alternative Agriculture News* (monthly newsletter), and *American Journal of Alternative Agriculture* (quarterly).

**International Centre for Genetic Engineering and Biotechnology**
Padriciano 99
I-34012 Trieste, Italy
39 (40) 37571
Fax: 39 (40) 226555
Telex: 460396 ICGEBT I
Contact: Director

This multinational organization, founded in 1984, promotes the use of biotechnology to solve problems of developing nations. It sponsors training programs and research and development in health care and agriculture, as well as industrial applications of biotechnology. The center advises on biotechnology issues and maintains a reference library, a database, and electronic mail.

*Publications: Activity Report* (annual), *Helix* (quarterly newsletter in English).

**International Plant Biotech Network**
c/o TCCP, Colorado State University

Dept. of Biology
Fort Collins, CO 80523
(303) 491-6996
Fax: (303) 491-2293
Telex: 3711418
Contact: Director

The International Plant Biotech Network, founded in 1985, connects researchers and interested individuals promoting the use of plant biotechnology to improve crops, particularly in developing nations. It fosters the exchange of information among researchers in plant tissue culture and genetic engineering and provides a consulting service for individuals and organizations. The organization maintains a database on plant biotechnology.

*Publication: International Plant Biotechnology Network Newsletter* (semiannual).

**March of Dimes Birth Defects Foundation**
1275 Mamaroneck Avenue
White Plains, NY 10605
(914) 428-7100
Fax: (914) 428-8203
Contact: President

Founded in 1938 by President Franklin D. Roosevelt, the March of Dimes promotes prevention of birth defects via proper prenatal care by addressing mother–child health issues such as low birthweight and maternal substance abuse.

*Publication: Genetics in Practice* (quarterly newsletter). A website (http://www.modimes.org) provides educational material on genetic diseases, including spina bifida, sickle cell disease, the neurofribramatoses, cleft lip and palate, clubfoot, and Down's syndrome.

**Michael Fund (International Foundation
for Genetic Research)**
500A Garden City Drive
Monroeville, PA 15146-1128
(412) 823-6380
Contact: Director

This fund, with its prolife stance, has supported research on Down's syndrome and related genetic disorders since 1978. The group opposes prebirth detection of disorders as well as abortion or euthanasia of afflicted children or adults. As an alternative it seeks to improve care, treatment, and rehabilitation of children and adults through the support of professionals and public advocates.

*Publications:* Newsletter; website (http://www.cvzoom.net/~chuckdet/index.html).

**N. I. Vavilov All Union Society of Geneticists and Selectionists**
Ulitsa Fersmana 11, Building 2
117312 Moscow, Russia
(95) 124-5952
Contact: President

Founded in 1966, the Society promotes development in all fields of genetics and selection in the former USSR. It maintains a reference library and supports educational programs concerned with biodiversity, protection of the gene pool, and informs the general public about genetic issues.

*Publication: Handbook* (periodic).

**National Council on Gene Resources**
1738 Thousand Oaks Blvd.
Berkeley, CA 94707
(510) 524-8973
Contact: Executive director

The council has linked organizations and individuals interested in the use, management, and conservation of the world's genetic resources since 1980. It is concerned with the loss of genetic diversity and its impact on the economy and on quality of life in the agricultural, forestry, pharmaceutical, and fishing industries. In addition, it produces and sponsors educational materials and programs for the general public.

*Committees:* Anadromous Salmonid Gene Resources, Barley Gene Resources, Douglas Fir Gene Resources, Strawberry Gene Resources. *Program: California Gene Resources.*

**National Foundation for Jewish Genetic Diseases**
250 Park Avenue, Suite 1000
New York, NY 10177
(212) 371-1030
Contact: President

The foundation, established in 1974, seeks to have an impact on the seven known genetic diseases affecting children of mainly

Ashkenazi Jewish heritage (Gaucher disease, dysautonomia, Tay-Sachs disease, Bloom syndrome, Niemann-Pick disease, and mucolipidosis IV). It supports basic medical research and acts as a referral agency for individuals. The organization promotes the establishment of disease carrier identification, provides genetic counseling, including prenatal detection, and conducts a national education campaign.

*Publications:* A variety of books and brochures on Tay-Sachs, ß-lipoproteinemia, Bloom Syndrome, Canavan Disease, Familial dysautomia, Gaucher's Disease, Mucolipidosis IV, Niemann-Pick, and Torsion distonia.

### National Fragile X Foundation
1441 York Street, Suite 303
Denver, CO 80206
(303) 333-6155; (800) 688-8765
Fax: (303) 333-4369
Contact: Executive director

Founded in 1974, the foundation is a multinational organization uniting professionals and families to support research and to improve the treatment of this X chromosome–linked genetic disease that causes mental retardation in affected males. It supports, advises, and assists parents of children with fragile X syndrome. The organization maintains a reference library.

*Publications: National Fragile X Foundation Newsletter* (bimonthly), audio and videotapes, brochures and information packets for families and professionals.

### National Health Federation
P.O. Box 688
Monrovia, CA 91017
(818) 357-2181
Fax: (818) 303-0642
Contact: President

This large, established, multinational organization, founded in 1955, seeks alternatives to "organized medicine, the pharmaceutical industry, and other special interests" in support of individual freedom of choice in matters of health. The group serves as watchdog and attempts "corrective" measures through education, legislation, and coordination of like-minded organizational efforts. The federation supports research in such areas as laetrile testing and conducts legislative lobbying. It maintains a computer database and a public library of some 25,000 entries on alternative health and related laws.

*Publication: Health Freedom News* (periodic).

**National Institute for Science, Law, and Public Policy**
1424 16th St. NW, Suite 105
Washington, DC 20036
(202) 462-8800
Fax: (202) 265-6564
Contact: President

The institute was founded in 1978 to influence public policies on sustainable agriculture, food safety, and nutrition. It promotes technologies avoiding chemical fertilizers and pesticides. The organization operates an information clearinghouse on aspartame, federal regulatory practices on milk pricing, use of prescription drugs, and interpretation of food and drug law. Training is provided through internships in areas of concern. The group provides information on request and maintains a speakers bureau.

*Publications: The Earth and You, Eating for Two, Eat Wise,* and *Healthy Harvest IV: A Directory of Sustainable Agriculture Organizations* (annual).

**National Neimann-Pick Disease Foundation**
22201 Riverpoint Trail
Carrollton, VA 23314
(804) 357-6774
Fax: (804) 357-6774
Contact: Chairman

This multinational organization of parents, friends, and health and education professionals, founded in 1992, promotes research on Niemann-Pick disease. This genetic disease affects sphingomyelin lipid and cholesterol metabolism. The foundation provides medical and educational information, offers support to families of children with the disease, supports legislation, and sponsors research on Niemann-Pick disease. It also maintains a reference library and a telephone referral service.

*Publication: Niemann-Pick Newsletter* (quarterly).

**National Society of Genetic Counselors**
233 Canterbury Drive
Wallingford, PA 19086

(610) 872-7608
Fax: (610) 872-1192
E-mail: beansgc@aol.com
Contact: Executive director

Founded in 1979, this multinational professional organization is composed of genetic counselors and people with an interest in genetic counseling. A major focus is on education of the genetic counselor and promotion of the field in view of the increased need for such services with the advent of new genetic tests. The group provides a speaker's bureau and compiles statistics related to genetic counseling. It maintains a computer job matching service and a general database.

*Publications: Journal of Genetic Counseling* (quarterly), *Perspectives in Genetic Counseling* (quarterly), a variety of high school and college level career packets.

**Neurofibromatosis Foundation, Inc.**
8855 Annapolis Road, Suite 110
Lanham, MD 20706-2924
(301) 577-8984; (800) 942-6825
TDD: (410) 461-5213
Fax: (301) 577-0016
Contact: President

A multinational support group founded in 1988 to serve individuals with neurofibromatosis, their parents, caregivers, and health care providers. Neurofibromatosis is a genetic disorder linked to a number of neurologic disabilities in which tumors form on nerves. The organization informs local, state, and national legislators about the needs of neurofibromatosis families. It promotes, supports, and funds medical and sociological research, including an annual scholarship, on possible cures and treatments. Neurofibromatosis Foundation identifies peer-counseling and local support resources, and offers referrals to medical resources. It maintains an archival library that is not, however, open to the public.

*Publications: Neurofibromatosis Ink* (newsletter, periodic), *Understanding Neurofibromatosis: An Introduction for Patients and Parents* (booklet).

**Physicians' Committee for Responsible Medicine**
P.O. Box 6322
Washington, DC 20015
(202) 686-2210

Fax: (202) 686-2216
Contact: President

This committee is a large lay organization of physicians, scientists, health care professionals, and interested others seeking to increase awareness of the importance of preventive medicine and nutrition. Founded in 1985, the group raises scientific and ethical questions about the use of animals and humans in medical research. Recent activity includes organizing a coalition of environmental and public health groups to push for adoption of a requirement for short-term nonanimal tests in lieu of those regulations mandating animal tests. The Physicians Committee maintains a speakers bureau and both general and breast cancer hotlines.

*Publications: Alternatives in Medical Education, Food for Life, Good Medicine* (quarterly), *The Power of Your Plate*, brochures and fact sheets.

**Prader-Willi Syndrome Association (U.S.A. Branch)**
2510 S. Brentwood Blvd., Suite 220
St. Louis, MO 63144
(314) 962-7644; (800) 926-4797
Fax: (314) 962-7869
Contact: Executive director

The association is a multinational of families and professionals, founded in 1975, that works to promote communication about this genetic syndrome and how to cope with it. It supports research and establishment of treatment facilities, conducts educational programs, and compiles statistics. The group maintains a reference library and a computer database.

*Publications: Gathered View* (bimonthly newsletter), *Management of Prader-Willi Syndrome* (book), and various information packets.

**Public Responsibility in Medicine and Research**
132 Boylston Street, 4th floor
Boston, MA 02116
(617) 423-4112
Fax: (617) 423-1185
Contact: Executive director

This is a multinational group of researchers, clinicians, administrators, attorneys, and laypersons supporting responsible animal and human research. The organization was founded in 1974 and is allied with the Applied Research Ethics National Association

(ARENA) but targeted more toward general education. It educates the health care community about development of research regulation and attempts to constructively counter the public's increasingly hostile attitude toward scientific research. The group sponsors public forums for discussion of issues, and acts as a resource for members in preparing presentations for legislative or other types of public hearings.

*Publications: Conference Report* (semiannual proceedings), *Guidebook on Institutional Animal Care and Use Committees, Human Subjects Guidebook.*

**The Pure Food Campaign**
1660 L Street NW #216
Washington, D.C. 20036
(202) 775-1132; (218) 225-4164
Fax: (218) 226-4157
E-mail: purefood@mr.net
Website: http://www.interactivism.com/purefood/
Contact: Director

The Pure Food Campaign is a nonprofit organization dedicated to sustainable methods of food production and consumption in the U.S. and the world. It is sponsored by the Washington, D.C.–based Foundation on Economic Trends whose president is the technology critic Jeremy Rifkin. It is an activist organization opposing genetic engineering, life form patenting, and chemical treatments of foods. Publications, online resources, boycott information, activist networking, and media events are available to journalists, organizations, and individuals.

**Rural Advancement Fund International–U.S.A. (RAFI–USA)**
P.O. Box 655
Pittsboro, NC 27312
(919) 542-1396
Fax: (919) 542-0069
Contact: Executive director

This multinational organization, founded in 1990, promotes conservation and the sustainable use of agriculture, family farms, and socially responsible uses of new technologies.

*Publications: The Community Seed Bank Kit* (preserving traditional crop varieties), *The Laws of Life: Another Development and the New Biotechnologies* (social and economic impact of new biotechnologies on the Third World), *RAFI Communique* (issues of biodiversity, biotechnology, and intellectual property), *Shattering: Food, Politics, and the Loss of Genetic Diversity* (governments' and corporations' struggle for control of access to the world's plant genetic resources).

**Society in Opposition to Human-Animal Hybridization**
23 Alabama Avenue
Lake Hopatcong, NJ 07849
(201) 663-4334
Contact: Thomas W. Chittum, founder

The society is a small, multinational organization founded in 1993 and composed of individuals opposed to human-animal hybridization on moral and ethical grounds. It seeks a worldwide ban on the creation of any and all human-animal hybrids.

*Publications:* Monthly newsletter.

**Turner Syndrome Society of the U.S.**
15500 Wayzata Blvd., No. 811
Wayzata, MN 55391
(612) 475-9944; (800) 365-9944
Fax: (612) 475-9949
Contact: Executive officer

Founded in 1987, the society seeks to link individuals suffering from the disease, families, and health care professionals. Turner syndrome is a genetic disorder affecting females and resulting in short stature and kidney, cardiac, and motor perception difficulties. The group promotes public awareness of the medical and sociopsychological impact of the syndrome and provides health care professional referral. It maintains a speakers bureau, reference library, and a computer database.

*Publications: Newsletter* (quarterly), *Turner's Syndrome: A Guide for Families, Turner's Syndrome: A Guide for Physicians.*

**World Aquaculture Society**
Louisiana State University
143 J. M. Parker Coliseum
Baton Rouge, LA 70803
(504) 388-3137
Fax: (504) 388-3493
Contact: Manager

The society, founded in 1970, is a multinational organization dedicated to the evaluation, promotion, and distribution of scientific and technological advancement in marine sciences worldwide. It promotes education and technical training and disseminates information on issues in aquaculture and mariculture.

*Publications: Advances in World Aquaculture* (periodic), *Journal of the WAS* (quarterly), *World Aquaculture* (quarterly; science, technology, political, and economic updates).

# Selected Print Resources 6

This chapter provides a selected annotated bibliography of printed resources (books, reports, periodicals, and directories) on different aspects of genetic engineering. The books are grouped according to their main subject content—general (including history), ethics, international, legal, business, agriculture, environment, and science—and then arranged alphabetically. Sources that disagree with the mainstream opinion are indicated with an asterisk. A listing of books on genetic engineering for young adults and younger readers is also provided. Following this are selections of U.S. government, congressional, agency, and nongovernment reports on applications of genetic engineering and its societal impact. A list of periodicals, newsletters, and directories provides an introduction to more current and often more focused information on scientific and social issues.

## Books

*An asterisk (\*) indicates sources with an alternative opinion to the mainstream.*

### General

Aldridge, S. *The Thread of Life. The Story of Genes and Genetic Engineering.* New

York: Cambridge University Press, 1996. 258 pp. ISBN 0-521-46542-7.

The focus of this book is primarily on the explanation of the cellular systems that molecular biologists have learned to control. By comparing the capabilities of humans to manipulate recombinant DNA with nature's evolutionary and contemporary processes, the author provides a useful perspective on their relative magnitudes. A section on how cellular DNA is organized in the genome and how that genetic information is regulated, as it is presently understood, gives an inkling of the complex program that the Human Genome Project hopes to elucidate. The relationship between genes and cancer provides a description of genetic predisposition that can be extrapolated to other diseases. An illuminating discussion of what a person's genetic identity means reveals that it differs from the public conception. Of importance for the educated lay reader, this particular section exposes the myth, often inferred in the media, that because certain simple things have been done, complete human or other organism engineering will be a trivial operation. Not only society but Nature conspires against genetic engineers. Examples are discussed in some detail of the uses to which biotechnology has been put in biopharmaceuticals, medicines, agriculture, and industrial production, as well as in environmental remediation and energy production. The author leaves to the many other books a detailed treatment of ethical issues and the social impact of genetic engineering.

Bains, W. *Biotechnology from A to Z.* New York: Oxford University Press, 1993. 358 pp. ISBN 0-199-63334-7.

A detailed introduction to biotechnology and its applications is provided at a level requiring, for a full understanding, more technical experience than *The Thread of Life* by Susan Aldridge (see above). The potential consequences of application of the technology are also discussed in some detail. Genetic engineering, particularly as reflected in human cloning and gene therapy, and the numerous moral, ethical, and social issues that accompany these technologies receive less coverage, partly since these considerations have become prominent only after 1993.

Cherfas, J. *Man-Made Life. An Overview of the Science, Technology, and Commerce of Genetic Engineering.* New York: Pantheon Books, 1982. 270 pp. ISBN 0-394-52926-X.

This book recounts the history of the science of genetic engineering. It is conversationally written and identifies many of the scientists involved. The author provides lucid explanations of what the scientists were doing and why, during the development of the technology rather than dryly presenting the facts.

Department of Government Relations and Science Policy. *Biotechnology.* Information pamphlet. Washington, DC: American Chemical Society, 1995. 16 pp.

A basic primer on biotechnology and its applications, compressed into a few pages.

DeSalle, R., and D. Lindley. *The Science of Jurassic Park and the Lost World or, How to Build a Dinosaur.* New York: Basic Books, 1997. 194 pp. ISBN 0-465-07379-4.

*Jurassic Park* and *The Lost World*, those fantastic, thrilling, Spielberg-directed movies with dinosaurs recreated from preserved DNA by recombinant DNA technology, were a marvel of special effects and cinematic art. The story lines of Michael Crichton's books on which these movies are based are similarly a marvel of science and science fiction, highly entertaining and seemingly believable. In the tradition of the great science fiction author Jules Verne, there is just enough fact to give the illusion of reality. Before getting too carried away about the possibilities of cloning ancient life, it's worth considering what the chances are that such events could really happen. *The Science of Jurassic Park* is a highly readable assessment of what series of unlikely events would have to occur, what would be required to bring them about, and under what set of conditions. Along the way the reader finds out quite a lot about recombinant DNA technology, its strengths and its weaknesses, the reality of cloning technology, the likelihood of dinosaur societies, and survival of disease. Finally the author considers the ethical questions of resurrecting an extinct creature. Even if it could be done, should it?

Fox, M.W. *Superpigs and Wondercorn: The Brave New World of Biotechnology and Where It All May Lead.* New York: Lyons and Burford, 1992. 209 pp. ISBN 1-558-21182-9.

A critical assessment of uncontrolled use of biotechnology and genetic engineering. While not opposed in principle to its use, the author warns of the overselling of the possibilities. The text is liberally sprinkled with facts and figures supporting the theme of the book.

Hodgson, J. G. *bio-Technology: Changing the Way Nature Works.* London: Cassell Publishing, 1989. 124 pp. ISBN 0-304-31783-7.

Although somewhat outdated and lacking coverage of the newest trends, this book provides a first-rate introduction to biotechnology and its applications. Lavishly illustrated and filled with color photographs and computer models of the technology and the people at work in the field, this book is a good place to get a visual feel and an understanding for the science.

*Hubbard, R., and E. Wald. *Exploding the Gene Myth: How Genetic Information Is Produced and Manipulated by Scientists, Physicians, Employers, Insurance Companies, Educators, and Law Enforcers.* Boston: Beacon Press, 1993. 206 pp. ISBN 0-807-00418-9.

As the book title suggests, Dr. Ruth Hubbard has a lot to say about the influence the genetic engineering revolution has or could have on society. The populist sentiment is evident, decrying the abuse of genetic information and the seemingly conspiratorial machinations of medicine, science, government, and industry combining to oppress certain individuals and minority groups. The number of alleged improprieties and potential ulterior motives detailed suggests an incredible scandal. While there is probably a certain amount of nefarious activity, as there is in any human endeavor, only part of the story is told here and the reader is left to judge the extent of the truth. Reality likely lies somewhere in between.

The technical background of the controversy and genetic concepts are explained in a facile manner. Despite having to impart a great deal of information, Dr. Hubbard presents her point of view in an engaging style aimed at the lay public. The book's notes and information resources give access to points of view often at odds with the scientific and government mainstream.

Judson, H. F. *The Eighth Day of Creation. Makers of the Revolution in Biology.* Plainview, NY: Cold Spring Harbor Press, 1996. 714 pp. ISBN 0-879-69477-7.

This is a history of molecular biology that concentrates less on the science and more on the political and social repercussions of the science. The previous edition of this book has not been available in the U.S. for a number of years. The new version updates events since 1978 and provides additional material on some of the principal historical figures.

*Kimbrell, A. *The Human Body Shop. The Engineering and Marketing of Life*. San Francisco: Harper, 1993. 348 pp. ISBN 0-062-50524-6.

This book takes an antitechnology view of the developments in organ transplant technology and genetic engineering. The author deals with a number of controversial issues concerning the patenting of life, use of body parts, and changing notions of self. Many of the arguments are made on the basis of what is "natural."

Kolata, G. B. *Clone. The Road to Dolly, and the Path Ahead*. New York: William Morrow and Company, 1998. 276 pp. ISBN 0-68815-692-4.

The author is the New York Times reporter who broke the story of the sheep Dolly, the first mammal cloned from differentiated tissue cells, to the American public. In this book she presents her case in the style of scientific journalism rather than as a philosopher or as a self-proclaimed moralist. The history of the science follows the step-by-step process of the development of the technology through the widely accepted technology of cloned cattle embryos of prized stock to the virtually ignored creation of the twin lambs Megan and Morag from cloned embryonic cells, up to the highly publicized cloning of Dolly from a differentiated adult udder cell. Ms. Kolata quotes a variety of opinions and in the end raises more questions than she answers. In the meantime she explains clearly and understandably to a wide nontechnical audience how and why society has come to this point in the cloning issue. She exposes some of the future implications of the technology and our reaction to it, for further contemplation.

Krimsky, S. *Genetic Alchemy: The Social History of the Recombinant DNA Controversy*. Cambridge, MA: MIT Press, 1982. 445 pp. ISBN 0-262-11083-0.

A history of the early years of the development of recombinant DNA technology.

*Lear, J. **Recombinant DNA: The Untold Story.** New York: Crown Publishers, 1978. 280 pp. ISBN 0-517-53165-8.

John Lear, a veteran commentator on the communication between science and society, presents an exposé-style view of the history of the recombinant DNA controversy. He weaves an engrossing tale of personalities and motives in which he detects a conspiracy of scientists to deceive and control the public. This highly readable book, written on a Ford Foundation grant, presents an alternative critical picture of how science and social issues interact. It is particularly instructive to consider these thoughts after the more than twenty years of genetic engineering practice that have passed since the publication of this book. Just what have society and scientists learned about each other in that time?

Levine, J. S., and D. Suzuki. *The Secret of Life: Redesigning the Living World.* Boston: WGBH Television, 1993. 280 pp. ISBN 0-963-68810-3.

The book form of the popular "Nova" educational television series on genetic engineering.

Lyon, J., and P. Gorner. *Altered Fates.* New York: W.W. Norton and Company, 1995. 636 pp. ISBN 0-393-31528-2.

Written by two *Chicago Tribune* journalists who won the Pulitzer Prize for reporting in 1987, this book provides an enlightening and entertaining treatment of the history of gene therapy and the development of the Human Genome Project. It includes discussions of gene repair and work with embryos. After making the case that genetic diseases have a great impact, the authors make it clear that they believe society is still not ready to handle the implications of gene therapy.

Paula, L., editor. *Biotechnology for Nonspecialists—A Handbook of Information Sources.* EFB Task Group on Public Perceptions of Biotechnology, 1997. 286 pp. ISBN 9-07611-001-8.

This handbook is designed to assist people with a broad range of interests and backgrounds in finding information on the public debate about biotechnology in Europe. Written information sources are cataloged, and lists and descriptions of organizations are provided, along with a collection of Internet sites and acronyms. The book is intended to be an information source and organizer rather than to provide detailed scientific explanations the author feels are found elsewhere.

Piller, C., and K. R. Yamamoto. *Gene Wars: Military Control over the New Genetic Technologies.* New York: Beech Tree Books, 1988. 302 pp. ISBN 0-688-07050-7.

A somewhat dated account of the involvement of the military with genetic engineering and recombinant DNA technologies.

*Rifkin, J. *The Biotech Century. Harnessing the Gene and Remaking the World.* New York: Jeremy P. Tarch/Putnam, 1998. 271 pp. ISBN 0-874-77909-X.

As he suggests in the title, Jeremy Rifkin has a few comments to make about the genetic engineering revolution that he opposed more than twenty years ago. The specter of human cloning (by nuclear transfer) has reawakened the original concerns of the early anti-genetic engineering activists. Along the way Rifkin reaffirms his Luddite leanings about the uses and abuses of technology in his examples of the acquisition of fire, the early printing press, and the English law on common land. He finally passes on to what he perceives as our failing modern industrial workplace. His message—to consider the ethical quandaries of technology in general, and those of biotechnology and genetic engineering in particular—comes through as it did in *Algeny,* although with less force and alarm than in the past.

*Rifkin, J., and N. Perlas. *Algeny.* New York: Viking Press, 1983. 298 pp. ISBN 0-670-10885-5.

In this book, Rifkin presents a critique of the then emergent biotechnology epoch. The term algeny, newly coined by the author, is drawn in parallel to alchemy. In the ancient world alchemy was the culmination of the natural conversion of lesser materials into gold or perfection through the agency of fire. Biotechnology is cast as an analogous philosophy in which technology provides the molding force for change to the detriment of natural systems (spiced throughout by Rifkin with a hint of conspiracy). This world view in which technology is destined to destroy the natural order of the relationship of humankind to the earth is the driving force of the author's actions in opposition to biotechnology and his many publications on ecological, social, and political issues.

Russo, E., and D. Cove. *Genetic Engineering: Dreams and Nightmares.* New York: W.H. Freeman/Spektrum, 1995. 243 pp. ISBN 0-716-74546-1.

A detailed lay presentation of the promises and disasters of genetic engineering.

*Shiva, V., and I. Moser. *Biopolitics: A Feminist and Ecological Reader on Biotechnology.* Atlantic Highlands, NJ: Zed Books, 1995. 294 pp. ISBN 1-856-49335-0.

A nontraditional view of biotechnology.

Thompson, L. *Correcting the Code: Inventing the Genetic Cure for the Human Body.* New York: Simon and Schuster, 1994. 378 pp. ISBN 0-671-77082-9.

Larry Thompson, a master's-level molecular biologist and a former science writer for *The Washington Post*, tells the history of genetic therapy, from the initial grandiose plans to the realization that developing the process would be long and hard, and finally to the first true genetic treatment of the four-year-old Ashanti DeSilva in September 1990. He vividly portrays the very personal stake the genetic pioneers had in their science, and the disappointments and temptations, set in the background of the early days of the genetic revolution.

Watson, J. D. *The Double Helix.* New York: W.W. Norton, 1980. 298 pp. ISBN 0-393-01245-X.

The controversial story of the discovery of the helical structure of DNA as seen through the eyes of James Watson, Nobel laureate. Watson shatters the image of the white-coated, calculating scientist toiling alone in a laboratory, with his account of the coincidences, the emotions, and the conflicts of strong personalities with different motives. Accounts by others involved in the search for the "Holy Grail of genetics" show the personal nature of science from diverse perspectives.

Watson, J. D., and J. Tooze. *The DNA Story. A Documentary History of Gene Cloning.* San Francisco: W.H. Freeman, 1981. 605 pp. ISBN 0-7167-1292-X.

This engaging history of the early days of the recombinant DNA story provides insight into the scientific, political, and legislative developments during that period. Reproductions of many letters exchanged between key scientists, administrators, and legislators bring this era to life by giving a view of the process rather than merely the final result. Reproductions of key portions of government documents, newspapers, magazines, and science journals provide the societal background throughout.

# Ethics

Andrews, L. B., J. E. Fullarton, N. A. Holtman, and A. G. Motulsky, editors. *Assessing Genetic Risks: Implications for Health and Social Policy.* Washington, DC: National Academy Press, 1994. 338 pp. ISBN 0-309-04798-6.

This is a report by the Institute of Medicine gleaned from a series of workshops and meetings, public and private, on genetic testing and its impact on patients, providers, and the laboratories providing the services. Each section concludes with a series of conclusions and the recommendations of the Committee on Assessing Genetic Risks. A unique contribution of this book is that it considers what is required on a practical level to achieve the accuracy and accountability of the laboratory science (chapter 3), public education to allow informed decision making (chapter 5), and training the genetic professionals who will be presenting the options (chapter 6). It also considers the usual social, ethical, and legal implications of genetics policies. This is an informative but densely written book.

Baker, Catherine. *Your Genes, Your Choices. Exploring the Issues Raised by Genetic Research.* Washington, DC: American Association for the Advancement of Science Directorate for Educational Resources, 1998. 82 pp. No ISBN.

This book is available on the World Wide Web at http://ehrweb. aaas.org/ehr/books/index.html. It is offered as an educational resource on genetic engineering technology and issues. Aimed at the lay public, it explains commonly used genetic terminology and provides examples of circumstances when people actually come into contact with the issues—health decisions, law enforcement, food, and reproductive choices. A concise introduction to the social impact of genetic engineering.

Bodner, W., and R. McKie. *The Book of Man: The Human Genome Project and the Quest to Discover Our Genetic Heritage.* New York: Simon and Schuster, 1995. 272 pp. ISBN 0-195-11487-6.

A description of the Human Genome Project and its implications for society.

Cole-Turner, Ronald. *The New Genesis. Theology and the Genetic Revolution.* Louisville, KY: Westminster/John Knox Press, 1993. 128 pp. ISBN 0-664-25406-3.

The author discusses genetic engineering and its uses in the context of humankind and Christian theology. He notes that in general, except for certain groups who oppose any technological intervention, little resistance is shown to medical applications. More concern is shown for any attempt to "improve" humans, although he acknowledges that medical necessity and "improvement" can converge on occasion.

Drlica, K. A. *Double-Edged Sword. The Promises and Risks of the Genetic Revolution.* Reading, MA: Addison-Wesley Longman, 1994. 256 pp. ISBN 0-201-40838-4.

The author concentrates primarily on the social and ethical consequences of making use of the influx of genetic information. He provides a list of genetic services providers and volunteer services for different genetic diseases.

Frankel, M. S., and A. Teich, editors. *The Genetic Frontier: Ethics, Law, and Policy.* Washington, DC: American Association for the Advancement of Science, AAAS Publication # 93-27S1994, 240 pp. ISBN 0-87168-526-4.

This collection of articles on the implications of the Genome Project and genetic testing is distilled from an invitational conference of the AAAS on social aspects of genetic engineering. Contributors comment on defining the family, privacy, linking genetics and behavior, assigning responsibility, use of genetics in criminal justice, and intellectual property and patent rights versus human dignity.

Holdrege, C. *Genetics and the Manipulation of Life: The Forgotten Factor of Context.* Hudson, NY: Lindisfarne Press, 1996. 190 pp. ISBN 0-940-26277-0.

A philosophical discussion of genes and organisms.

*Howard, T., and J. Rifkin. *Who Should Play God? The Artificial Creation of Life and What It Means for the Future of the Human Race.* New York: Delacorte Press, 1977. 272 pp. ISBN 0-440-09552-2.

This book considers the issues surrounding the increased control of genetics made feasible by recombinant DNA technology. The history of eugenics is recounted along with a description of modern reproductive technology and the impact the new technologies will have on social mores. Society sees recombinant DNA technology as a way to detect genetic defects and potentially to offer genetic fixes such as gene therapies. The authors pose the question whether these technologies are really needed or wanted by society. Examples of scientific/corporate abuses of power and mismanagement in other situations are cited to suggest that these should not be the groups that society allows to make decisions about using the new genetic technologies. At the time they wrote the book the authors were codirectors of the People's Bicentennial Commission. J. Rifkin is a well-known figure often found opposing the application of genetic technologies and involved with environmental issues.

Kevles, D. J., and L. Hood, editors. *The Code of Codes. Scientific and Social Issues in the Human Genome Project.* Cambridge, MA: Harvard University Press, 1992. 384 pp. ISBN 0-674-13645-4.

This collection of expert commentaries covers the spectrum of technologies that are required for the Genome Project to succeed and recounts its history through 1992. It also addresses many of the social issues that need to be confronted in the use of that information. Genetic testing—when it should be applied and what it means for the individuals involved—is the main theme of the book.

Kitcher, P. *The Lives to Come: The Genetic Revolution and Human Possibilities.* New York: Simon and Schuster, 1996. 381 pp. ISBN 0-684-80055-1.

The book is primarily directed at ethical issues although some scientific issues are considered.

Lee, T. F. *Gene Future: The Promise and Perils of the New Biology.* New York: Plenum Press, 1993. 339 pp. ISBN 0-306-44509-3.

This book describes the many sides of the issues of the science and ethics of genetic engineering and the Human Genome Project.

Nelkin, D., and L. Tancredi. *Dangerous Diagnostics: The Social Power of Biological Information.* New York: Basic Books, 1989. 207 pp. ISBN 0-465-01573-5.

This book considers the consequences of biological testing and the effects of labeling individuals in various social contexts such as the workplace, courts of law, and in health care.

O'Neill, T., editor. *Biomedical Ethics. Opposing Viewpoints.* San Diego: Greenhaven Press, 1994. 312 pp. ISBN 1-56510-062-X.

Ethical issues surrounding the medical applications of high technology include some that are brought about by the genetic revolution.

Reiss, M. J., and R. Straughan. *Improving Nature. The Science and Ethics of Genetic Engineering.* Cambridge, England: Cambridge University Press, 1996. 288 pp. ISBN 0-521-45441-7.

A biologist and a moral ethicist consider the potential of the genetic engineering revolution and explain its pitfalls. This book supplies much needed discussion on moral and ethical questions about the various aspects and applications of biotechnology. Pro and con viewpoints to each question are presented (without a conclusion), accompanied by critical facts and figures and without vilifying either viewpoint. This evenhanded presentation emphasizes that there really are (at least) two sides to every issue. In doing so, such a format also points out how far apart the two sides remain. The authors emphasize the importance of public education in informing people to have meaningful discussions about biotechnology issues, and they distinguish between education and information. Since the authors, a biologist (MJR) and a moral philosopher (RS), are English, much of the information is from the European community. They provide an unusual viewpoint compared to most U.S. sources.

Rothstein, M. A., editor. *Genetic Secrets: Protecting Privacy and Confidentiality in the Genetic Era.* New Haven, CT: Yale University Press, 1997. 511 pp. ISBN 0-300-07251-1.

The chapters in this compendium provide a comprehensive assessment of the impact of the new genetic technology and its medical and nonmedical uses on personal privacy. The "gen-etiquette" controversy and how it is being played out in the societal arenas of health and law and order are discussed in separate chapters by a group of clinical, scientific, legal, and ethics experts. Particularly valuable is the inclusion of the international impact of the new technology and the issues being debated in countries other than the United States, with the focus on Europe. A working policy framework for protecting genetic information and personal privacy is proposed.

Smith II, George P. *Bioethics and the Law: Medical, Socio-Legal, and Philosophical Directions for a Brave New World.* Lanham, MD: University Press of America, 1993. 332 pp. ISBN 0-819-19177-9.

Bioethics is the point at which genetic engineering and biotechnology make contact with society's norms and expectations. The law attempts to put some consistent form to these sometimes vague guidelines through legislation that prescribes which actions may be taken and which are not allowed. This book provides a discussion of this process.

Suzuki, D., and P. Knudtson. *Genethics. The Clash between the New Genetics and Human Values.* Cambridge, MA: Harvard University Press, 1990. 372 pp. ISBN 0-674-34566-5.

This is a book written for the nonscientist. The television science communicator David Suzuki and his coauthor provide treatment of the science and technology, following up with a very readable discussion. The bulk of the book is concerned with the ethical implications of genetic technology. Topics covered include the blaming of aggressive behavior on chromosomes (XYY males), genetic screening, somatic and germ cell therapy, biological weapons, environmental damage to DNA, the value of genetic diversity, and the implications of crossing genetic boundaries with transgenic organisms. The authors raise numerous questions about the use of genetic technologies.

# International

Brock, M. V. *Biotechnology in Japan.* The Nissen Institute/Routledge Japanese Studies Series, New York: Routledge Press, 1989. 156 pp. ISBN 0-415-03495-7.

The policymaking process used in Japan is different from that in the U.S. In Japan, a coalition between government bureaucracy, industry, and politics is essential. Such close ties would engender a congressional investigation in the U.S. Japan is particularly attracted to biotechnology because it allows the efficient use of normally scarce resources in that island nation, thus reducing dependence on outside parties and economic forces. The shift to knowledge-intensive industries is significantly more advanced there than in the United States. Most of the book is concerned with strategies for developing a biotechnology industry in Japan.

Doyle, J. C., and G. J. Persley, editors. *Enabling the Safe Use of Biotechnology: Principles and Practice.* Washington, DC: World Bank, 1996. 84 pp. ISBN 0-8213-3671-1.

This book describes a global environment agenda whose stated goal is to provide "a practical guide for policymakers and research managers who are responsible for making decisions on ensuring the safe use of modern biotechnology." This report does not consider the global differences of opinion over introducing genetically engineered organisms into agriculture. This is the tenth in a series of monographs on "Environmentally Sustainable Development Studies."

Juma, C., J. Mugabe, and P. Kameri-Mbote, editors. *Coming to Life: Biotechnology in African Economic Recovery.* London: Zed Books, 1995. 192 pp. ISBN 9966-41-087-2.

The countries of Africa are experiencing economic, political, and ecological decline, all of which contribute to a decreasing ability to compete on the open world market. Their political and educational/industrial institutional structures are unable to cope with changes, such as a decreasing market for those raw materials that have constituted the bulk of the exports of these nations. The book assesses the situation in Cameroon, Ethiopia, Kenya, Tanzania, and Uganda. It makes a series of recommendations for African policymakers to improve the economic outlook for their region by facilitating the integration of new technology and institutional change.

Peritore, N. P., and A. K. Galve-Peritore. *Biotechnology in Latin America: Politics, Impacts, and Risks.* Wilmington, DE: SR Books, 1995. 229 pp. ISBN 0-8420-2556-1.

This collection of articles considers the general problems associated with the economies of developing nations and with their scientific and technical infrastructure. It specifically looks at the situation in Mexico, Cuba, and Columbia. Particularly relevant to these countries are the ideas of property rights (patent protection) balanced against gene pool rights (genetic property rights to natural diversity). Economic limitations for most Latin American nations have stunted the development of an appropriate infrastructure for technological advancement. This has led to the idea that some of that national debt might be forgiven in the form of credits for building debtor nations' biotechnological competence so that they would have the skills to boost their own economies, repay loans, and afford developed nations' products.

Russell, A. M. *The Biotechnology Revolution. An International Perspective.* New York: St. Martins Press, 1988. 266 pp. ISBN 0-7450-0013-4.

The emphasis of this book comes from the author's training in international relations. It provides a perspective on the process of development of regulations in the U.K. and Europe, in particular, with some comments on Japan, West Germany, and Canada. There is little on the U.S. The history of national developments and controls is reviewed. There is a fear that Western companies will test their genetic engineered products in underdeveloped countries, much as they sold less safe nuclear reactor designs there, because the less developed nations had less sophisticated regulatory standards.

Sasson, A. *Biotechnologies in Developing Countries: Present and Future. Vol. 1: Regional and National Survey.* Paris: UNESCO Publications, 1993. 764 pp. ISBN 9-331-02875-8.

Sasson has been the director of the Bureau of Studies, Programming, and Evaluation of UNESCO since 1974. This book provides a wealth of economic data, and information about biotechnological research and development activities. It presents an overview of achievements, expected developments, constraints on progress and application, research and industrial application strategies, consumer interests and biosafety, intellectual property, and regulation issues. The book concludes with comments on the economic impact (with a set of forecast tables) and on how biotechnology is being pursued in developing countries.

Smith, J. E. *Biotechnology.* 3rd edition. London: Cambridge University Press, 1996. 236 pp. ISBN 0-521-44467-5.

This book considers different aspects of genetic engineering particularly with respect to industrial applications. It covers the biotech industry, agriculture, biomining, concerns with patents and ethics, and environmental release of genetically modified organisms. The information is primarily from England, providing a distinctive European view of the subject.

## Legal

Billings, P. R., editor. *DNA on Trial: Genetic Identification and Criminal Justice.* Plainview, NY: Cold Spring Harbor Laboratory Press, 1992. 154 pp. ISBN 0-879-69379-7.

This book covers the history and impact of DNA forensic evidence on civil liberties and public policy. It contains an analysis of decisions made in different courts by juries considering DNA evidence.

Cook, T., C. Doyle, and D. Jabbari. *Pharmaceuticals, Biotechnology, and the Law.* New York: Stockton Press, 1991. 834 pp. ISBN 1-561-59040-1.

The laws respecting pharmaceuticals and biotechnology differ, sometimes significantly, between nations. This book reviews the law and legislation governing pharmaceuticals and biotechnology, concentrating on Great Britain and the European Economic Community countries. The appendices contain the texts of the EEC Council Directives and other related documents of the late 1980s to early 1990s.

Environmental Law Centre. *Law in the New Age of Biotechnology.* Edmonton, Alberta, Canada: Environmental Law Centre, 1992. 220 pp. ISBN 0-921-50341-5.

This book describes the legal basis and present legislation governing biotechnology and environmental law in Canada.

Inman, K. and Rudin, N. *An Introduction to Forensic DNA Analysis.* Boca Raton, FL: CRC Press, 1997. 256 pp. ISBN 0-849-38117-7.

A discussion of modern methods of analyzing DNA and the use of that information in courts of law. A technical source.

Shiva, V. *Biopiracy: The Plunder of Nature and Knowledge.* Boston: South End Press, 1997. 148 pp. ISBN 0-896-08556-2.

A confrontation is shaping up over the legal status and the growing economic importance of genetic diversity and products from underdeveloped countries. Pharmaceutical companies in particular, as well as companies with interests in agriculture, regard the untapped natural resources and folk medical practices of the developing nations as useful starting points for novel product development. Indigenous peoples are beginning to recognize the real value of what they had been giving away for so many years and have started to take ownership. The legal status of their proprietary position and the demand for free access to world natural resources for research purposes are the source of much discussion and legal wrangling.

Sibley, Kenneth D., editor. *The Law and Strategy of Biotechnology Patents.* Biotechnology Series No. 25. Boston: Butterworth-Heinemann, 1994. 262 pp. ISBN 0-750-69444-0.

This book focuses on the law governing biotechnology patents and the strategies followed in protecting intellectual property in the United States.

Weir, B. S., editor. *Human Identification: The Use of DNA Markers.* Contemporary Issues in Genetics and Evolution Series. Boston: Kluwer Academic Publishers, 1995. 213 pp. ISBN 0-792-33520-1.

This monograph explains the methodology and the interpretive issues involved in using DNA patterns to establish whether individual people are related to each other.

Wiegele, Thomas C. *Biotechnology and International Relations. The Political Dimensions.* Gainsville, FL: University of Florida Press, 1991. 212 pp. ISBN 0-8130-1055-1.

Global interdependence in science has created an international political context for biotechnology. This book is written to provide background for those persons pursuing a career in international relations. It tries to show where this particular expertise can assist governments in dealing with the new technology to maximize benefits and minimize harm. It reviews national regulatory postures, major international legal instruments, determination of harm and sanctions, and verification and enforcement of laws governing biotechnology applications. Other concerns are the impact of international commerce on both developed and Third World economies and the prospect for biological warfare as a terrorist weapon.

World Intellectual Property Organization. *Guide on the Licensing of Biotechnology.* Geneva: World Intellectual Property Organization, Publication No. 708(E), 1992. 197 pp. ISBN 9-280-50410-X.

The patenting of biotechnology products is a highly specialized branch of intellectual property rights. It is at the center of the agricultural commercialization controversy that is far from settled. This book provides a summary of those established concepts that are accepted, at least by the developed nations.

Wright, Susan. *Molecular Politics: Developing American and British Regulatory Policy for Genetic Engineering, 1972–1982.* Chicago: University of Chicago Press, 1994. 591 pp. ISBN 0-226-91065-2.

This book describes in considerable detail the history of the development of regulation of biotechnology and genetic engineering. The author outlines the cycle of regulation followed by deregulation, with an analysis of the societal, political, economic, and scientific forces that have driven events. She interprets deregulation as a sellout of the public due to overriding industrial (and research) interests rather than a reasoned change.

# Business

Kenney, M. *Biotechnology: The University-Industrial Complex.* New Haven, CT: Yale University Press, 1986. 306 pp. ISBN 0-300-03392-3.

This book describes the early development of biotechnology and the university-industrial connection from a sociological point of view. The extremely close relationship with industry has not been without its effects on the university. The structure of the biotech industry is put into perspective—venture capital funding, investments by pharmaceutical and chemical companies, and the rewards and incentives for academics. Biotech companies have undergone tremendous growth that has brought changes in the roles, motivations, and perceptions of their employees. Growing relationships with multinationals are forcing further changes in industry, particularly in those applied to agriculture.

Krimsky, S. *Biotechnics and Society: The Rise of Industrial Genetics.* New York: Praeger Publishers, 1991. 280 pp. ISBN 0-27593-853-X.

A history of the evolution of the industrial involvement in genetic engineering, from enhancing technology to commercialization of entirely new ideas and products.

van Balken, J.A.M., editor. *Biotechnological Innovations in Chemical Synthesis.* London: Butterworth-Heinemann, 1997. 376 pp. ISBN 0-750-60561-8.

This book describes the current and potential uses of biotechnology in chemical synthesis. Industrial fermentation processes are connected with their respective chemical syntheses to compare their strengths and weaknesses and where they might be applied together to advantage.

# Agriculture

*Busch, L., W. B . Lacy, J. Burkehardt, and L. R. Lacy. *Plants, Power, and Profit: Social, Economic, and Ethical Consequences of the New Biotechnologies.* Cambridge, MA: B. Blackwell, 1991. 275 pp. ISBN 1-557-86088-2.

The book discusses emerging issues and trends in biotechnology worldwide. It presents perspectives on science and society including a discussion of the philosophy behind the science. A history of plant breeding and its connection to the new genetic engineering technologies provides a background of the debate about the relative contributions of the two strategies. Examples of political biology with a chapter on wheat and one on tomatoes demonstrate the principles introduced in earlier sections. In a chapter on the international scope of the social impact of biotechnology, the authors demand accountability for releases of transgenic plants and for potential economic effects on indigenous industry. There is also the question of who will do the engineering on minority food crops and other crops important to many developing nations. With the highly technical nature of transgenic crops, the future in all countries appears to bode fuller integration of crop production into industry and removal of agriculture from the hands of individual farmers.

*Dawkins, Kristin. *Gene Wars. The Politics of Biotechnology.* Open Media Pamphlet Series. New York: Seven Stories Press, 1997. 60 pp. ISBN 1-888363-48-7.

This short book makes the case of the perceived perils of plant

and animal agricultural biotechnology, from the escape of engineered organisms to the impact on regular agriculture, such as
economic displacement and the evils of control exercised by the
multinational agricultural industry. The points are made succinctly and the book includes information sources to help activists make a difference.

*Gussow, J. D. *Chicken Little, Tomato Sauce, and Agriculture:
Who Will Produce Tomorrow's Food?* New York: The Bootstrap
Press, 1991. 150 pp. ISBN 0-94285-032-7.

A critical discussion of the impact of genetic engineering on food
production and on who will be producing food.

Institute of Food Technologies. *Appropriate Oversight for Plants
with Inherited Traits for Resistance to Pests.* Chicago: Institute
of Food Technologies, 1996.

This report recommends principles for regulating genetically engineered plants. The level of risk should be determined by the
characteristics of the plant, not lack of familiarity with the gene
change, source of genes, or method of gene transfer; the genes or
gene products should not themselves be pesticides that are subject
to federal control under the Insecticide, Fungicide, and Rodenti-
cide Act; concepts such as "generally regarded as safe" (GRAS)
applied to other areas such as food additives should also be applied to the new varieties of plants. GRAS substances are not subject to the federal Food, Drug, and Cosmetic Act. This report does
not include consideration of global differences of opinion over introducing genetically engineered organisms into agriculture.

Krimsky, S., and R. P. Wrubel. *Agricultural Biotechnology and
the Environment: Science, Policy, and Social Issues.* Urbana, IL:
University of Illinois Press, 1996. 294 pp. ISBN 0-252-02164-9.

This book describes the various applications of genetic engineering to agricultural systems. It covers herbicide-resistant plants,
insect-resistant plants, disease-resistant crops, various transgenic
plant products, microbial pesticides, nitrogen-fixation, and frost-
inhibiting bacteria. Animal agricultural products such as animal
growth hormones and various transgenic animals are included.
In keeping with his other publications, Krimsky is careful to include discussion of the cultural and symbolic sides of agricultural biotechnology.

*Mather, R. *Garden of Unearthly Delights. Bioengineering and the Future of Food.* New York: Penguin Books, 1995. 205 pp. ISBN 0-525-93864-8.

The author, the food editor of the *Detroit News*, considers the future of food supply in the United States through a series of visits with individuals involved in the process—from the new genetically engineered plants and recombinant bovine growth hormone for meat and milk products to sustainable agriculture. The tone of the commentary and the information selected for presentation do not favor the application of biotechnology to agriculture. The author suggests a blueprint for individual action in consumer buying power to support "whole food" production.

*Perlas, N. *Overcoming Illusions About Biotechnology.* Atlantic Highlands, NJ: Zed Books, 1994. 119 pp. ISBN 1-856-49303-2.

The author considers that the impact of genetic engineering and biotechnology on the environment, natural resources, society, and the family farm is largely negative. He describes sustainable agricultural methods as a solution and the control of biotechnology by legislative and populist means.

Reborn, P. *The Last Harvest: The Genetic Gamble that Threatens to Destroy American Agriculture.* Lincoln: University of Nebraska Press, 1995. 269 pp. ISBN 0-803-28962-6.

The closing of seed germ plasm repositories is exacerbating the loss of genetic diversity brought about by current plant breeding programs and by the cultivation of vast monocultures. The author points out the vulnerability of homogeneous crop plants to disease and to environmental pressures. These crops are denied periodic access to the heterogeneity of traits that might be offered by other varieties through free exchange with genetic material in the wild.

Rissler, J., and M. Mellon. *The Ecological Risks of Engineered Crops.* Cambridge, MA: MIT Press, 1996. 168 pp. ISBN 0-26218-171-1.

This monograph sponsored by the Union of Concerned Scientists points out potential risks associated with the widespread commercial cultivation of transgenic crop plants. Most current traits being modified are specific pest- and herbicide-resistance traits that benefit processors but don't enhance nutritional value. The original idea of making plants drought- or salt-resistant, nitrogen fixing, or higher yielding has faded with the realization that these are multigenic traits that are less amenable to transgenic manipulation. The authors propose a specific testing program to determine the risks that a given transgenic plant might become a weed or pass the transgene to wild populations.

# Environment

Chaudhry, G. R., editor. *Biological Degradation and Bioremediation of Toxic Chemicals.* Portland, OR: Dioscorides Press, 1994. 515 pp. ISBN 0-931-14627-5.

This book describes the vast microbial metabolic diversity and the quest to construct recombinant strains to combine pathways in the same organism to improve their effectiveness in bioremediation.

Prieels, A-M. *Development of an Environmental Bio-Industry: European Perceptions and Prospects.* Lanham, MD: Dublin Institute, by the European Foundation for the Improvement of Living and Working Conditions. Distributed by UNIPUB, 1993. 131 pp. ISBN 92-826-4691-2.

This report is compiled from responses (in 1990) of persons and organizations to the general questions: Can biotech contribute to an improvement in the state of the environment, and why is bioindustry not more developed in Europe? Specific topics include the pollution treatment bioindustry, the scientific and technological bases of the environmental bioindustry, acceptance of the development of applications in the environmental sector, biotechnology and sustainable development, and the political and regulatory framework of the environmental bioindustry. Many of the problems in Europe are the same as those encountered in the United States.

Witt, S. C. *Biotechnology, Microbes, and the Environment.* Briefbook® Series. San Francisco: Center for Science Information, 1990. 219 pp. ISBN 0-91200-503-3.

This book is designed to provide decision makers with a basic level of scientific understanding to make informed public policy as well as to educate the general public. Written in a distinctly irreverent but engaging style, the majority of the book is concerned with the impact of the technology, risk assessment, U.S. and international regulation, historical events, and the major issues at stake. A unique contribution is a list of U.S. and international "Expert Sources," along with comments on their areas of expertise. These experts are drawn from academic scientists, government agency workers, industrial representatives, and foes of genetic engineering. An assessment is offered of the "articulate interview" and the "solid source."

## Science

Bronzino, J. E., editor. *The Biomedical Engineering Handbook.* Boca Raton, FL: CRC Press, 1995. 2,862 pp. ISBN 0-849-38346-3.

A compendium on the design, development, and use of medical technology to diagnose and treat patients. It includes 11 chapters on biotechnology and 17 chapters on tissue engineering. Regulations and ethical principles currently in force that influence biomedical engineering are discussed.

Friedmann, T. *Gene Therapy: Fact and Fiction in Biology's New Approaches to Disease.* Plainview, NY: Cold Spring Harbor Laboratory Press, 1994. 124 pp. ISBN 0-879-69446-7.

This short book provides historical background and an assessment of current challenges. It reviews the field's technical achievements and ethical dilemmas.

Hall, S. S. *Invisible Frontiers: The Race to Synthesize a Human Gene.* New York: Atlantic Monthly Press, 1987. 334 pp. ISBN 0-871-13147-1.

The story of the race to clone and express the human insulin gene for the treatment of diabetes.

Jackson, J. F. *Genetics and You.* Totowa, NJ: Humana Press, 1996. 92 pp. ISBN 0-89603-329-5.

Written for the educated lay consumer of genetic testing, this short book explains the principles of genetics and how to make use of the test results in conjunction with genetic counseling. It clarifies the "new genetics" and provides guidance on how to get questions answered and on dealing with the decisions to be made.

Kreuzer, H., and A. Massey. *Recombinant DNA and Biotechnology: A Guide for Teachers and A Guide for Students.* American Society for Microbiology Press, 1996. 564 pp. ISBN 1-55581-101-9C (teacher), 364 pp.; ISBN 1-55581-110-8C (student).

An introductory textbook for high school and college students with textbook topics, wet lab and dry lab experiments and demos, and material on the impact on society—weighing risk and benefit, a decision making model for bioethical issues, case studies in bioethics, and careers in biotechnology.

Miklos, D. A., and G. A. Freyer. *DNA Science: A First Course in Recombinant DNA Technology.* Plainview, NY: Cold Spring Harbor Press, 1990. 477 pp. ISBN 0-89-278411-3.

This is a combination textbook/lab manual designed for advanced high school or beginning college students It shows where recombinant DNA technology came from and points where it might lead. It traces a historical perspective with experiments designed to illustrate the text. The course is designed to be supported by *Carolina Biological Supply* reagents and supplies. Well illustrated and engagingly written, the textbook itself is suitable for introducing courses on recombinant DNA in science and society.

Murray, T. H., M. A. Rothstein, and R. F. Murray, Jr. *The Human Genome Project and the Future of Health Care.* Bloomington: University of Indiana Press, 1996. 248 pp. ISBN 0-253-33213-3.

This book focuses on the implications of the Human Genome Project for society and health practice. Attention is drawn to the effect the information from the Genome Project will have on the quality and delivery of health care in the U.S. A series of articles by various experts addresses the issues of genetic discrimination in employment and insurance, equal access to testing and treatment for minority and indigent clients, and the impact of genome research on reproductive decision making.

Neel, J. V. *Physician to the Gene Pool. Genetic Lessons and Other Stories.* New York: John Wiley & Sons, 1994. 457 pp. ISBN 0-471-30844-7.

The author, a physician, discusses genetics and populations, genetics and individuals, and the impact of medical genetics on past and present medical care. He then compares these impacts to those expected from the future application of genetic engineering.

Rabinow, P. *Making PCR: A Story of Biotechnology.* Chicago: University of Chicago Press, 1996. 190 pp. ISBN 0-226-70146-8.

An entertaining story of the development of the polymerase chain reaction (PCR) method, a key technology in genetic engineering. The author provides a historical description of the events during the conception of this powerful tool for manipulating nucleic acids. The book engagingly includes numerous personal accounts of the struggle for power and precedence among the key players.

Svendsen, P., and J. Hau, editors. *Handbook of Laboratory Animal Science. Animal Models* (Volume 1). 224 pp. ISBN 0-849-34378-X. *Selection and Handling of Animals in Biomedical Research* (Volume 2). 448 pp. ISBN 0-849-34390-9. Boca Raton, FL: CRC Press, 1994.

A comprehensive description of lab animal genetics, diseases, health monitoring, nutrition, and environmental impact of the environment on animal testing. These books consider the ethics of animal experimentation in Europe and North America. Also discussed are alternatives to animal experiments, including isolated organs, cell cultures, and computer simulations.

# Books on Genetic Engineering (Young Adult Sources)

Balkwill, F. *Amazing Schemes Within Your Genes.* London: Harper-Collins, 1993. 32 pp. ISBN 0-001-96465-8.

The uniqueness of every human being is explained through the genes even though our DNA is 99.5% identical. The book explains and illustrates the process by which visible characteristics such as hair color, eye color, color of skin, and ear shape are transmitted. It shows how genetic diseases like cystic fibrosis are inherited. For ages 9–15.

Balkwill, F. *DNA is Here to Stay.* London: Harper-Collins, 1992. 32 pp. ISBN 0-001-91165-1.

The process by which the code of life directs the growth of an embryonic cell into a complete human being is described and clearly illustrated for ages 9–15.

Bornstein, S. *What Makes You What You Are: A First Look at Genetics.* Englewood Cliffs, NJ: J. Messner, 1989. 128 pp. ISBN 0-67168-650-X.

An introduction to genes, DNA, and genetics. For grades 7 and up.

Bryan, J. *Genetic Engineering.* New York: Thomson Learning, 1995. 64 pp. ISBN 1-56847-268-4.

The main topics of this book are the issues involved in the application of genetic engineering. The treatment is up-to-date, and the illustrations and pictures are very good.

Darling, D. J. *Genetic Engineering: Redrawing the Blueprint of Life.* Parsippany, NJ: Silver Barnett, Dillon Press, 1995. 64 pp. ISBN 0-87518-614-9.

This book outlines the progress made in understanding the genetic code by following research on cystic fibrosis. The author uses this disease, with its major effects on children, to explain mechanisms of genetic diseases and to enhance student involvement with the issues of genetic screening, amniocentesis, abortion, and gene therapy. For grades 5 and up.

Facklam, M., and H. Facklam. *From Cell to Clone. The Story of Genetic Engineering.* New York: Harcourt Brace Jovanovich, 1979. 128 pp. ISBN 0-152-30262-X.

This book cuts a wide swath, covering a little bit of everything—cell biology, genetics, DNA, cloning, and genetic engineering. The last chapter considers what it takes to become a scientist. The information is a bit dated, ca. 1979, but the concepts hold in essence.

Grace, E. S. *Biotechnology Unzipped: Promises and Realities.* Washington, DC: Joseph Henry Press (National Academy Press), 1997. 264 pp. ISBN 0-309-05777-9.

An up-to-date discussion of the possibilities and achievements of the new science of biotechnology.

Lampton, C. *DNA and the Creation of New Life.* New York: ARCO Publishing, 1983. 135 pp. ISBN 0-668-05396-8.

This somewhat dated book details the story of the quest for the gene and what scientists are doing to mold genes to do their bidding. The author, a well-known contributor to juvenile science literature, also considers some of the issues scientists face as the "engineers of life," including the question of genes as property. The ideas in Lampton's 1983 discussion about the projected uses and issues of DNA technology are very similar to those current now.

Lampton, C. *DNA Fingerprinting*. Danbury, CT: Franklin Watts, 1991. 112 pp. ISBN 0-531-13003-7.

An explanation of how DNA patterns in cells and body fluids are much like a fingerprint and how they can be used to identify individuals. Evidence based on DNA analysis can be used in a court of law to place an individual at the scene of a crime. For grades 7–12.

Marshall, E. L. *The Human Genome Project. Cracking the Code Within Us*. Danbury, CT: Franklin Watts, 1996. 128 pp. ISBN 0-531-11299-3.

The 15-year, multibillion dollar project of mapping and sequencing the human genetic code, the genome, is described and explained in this work for grades 8–12. The author explains the lay concepts involved in genetic research by concentrating on individual real scientists and the work they do as part of the project. The controversies involved in doing genetic research and in applying it are recognized, and the potential benefits of the knowledge for research and human health are once again brought to life through the device of telling stories about the people involved. Personalizing in this way helps to drive home the magnitude of the Human Genome Project. It also serves as a reminder that in the end it all comes down to individual people, the public, who will have to decide how to use the information.

Marteau, T., and M. Richards, editors. *The Troubled Helix: Social and Psychological Implications of the New Human Genetics*. Cambridge, England: Cambridge University Press, 1996. 359 pp. ISBN 0-521-46288-6.

This book addresses the interpretation and use of genetic information both at the personal level and by society as a whole. It has chapters on carrier testing of adults, prenatal testing, testing children, and the impact of genetic counseling. As the authors, pro and con, of the various chapters point out, disagreement rests on the lack of predictability of the outcome of having a genetic defect. While the technology for determining that there is a mutation in a specific gene has improved immensely, being able to predict when, where, and how much of an effect that lesion will have on an individual has not progressed appreciably. The interpretation problem will only get worse as geneticists tie together more complex multigenic traits. *The Troubled Helix* also analyzes social aspects of genetic testing. How do people react to and use genetic information to make decisions? Many people who are at risk for a devastating disease simply do not wish to know. Other points of debate include whether to test children and how and when to inform other members of a kindred about test results. There is, throughout, public agreement that genetic information is private for the individual. Numerous studies and statistics of various types are cited. The book is written from the English point of view, although a great deal of effort is made to make things more international. The reader will note that certain basic assumptions about public attitudes and legal precedence differ from those in the United States.

Sherrow, V. *James Watson and Francis Crick. Decoding the Secret of DNA.* Woodbridge, CT: Blackbirch Press, 1995. 110 pp. ISBN 1-56711-133-5.

This historical and biographical account of the discovery of the structure of DNA is written for the middle school or early high school student. It contains numerous illustrations and photographs of the important principles and of the human participants in the events leading up to the determination of the three dimensional structure of the DNA molecule. The story of Watson, Crick, and DNA is carried up through the 1990s, including Watson's brief tenure as head of the Human Genome Project.

Swisher, C. *Genetic Engineering.* San Diego: Lucent Books, 1996. 128 pp. ISBN 1-56006-179-0.

This book provides a comprehensive overview of genetic engineering technology placed in the perspective of the controversy brought by its applications. Opposing points of view are presented with equal weight to those of the proponents. A glossary, extensive index, list of organizations to contact for additional information, and suggestions for further reading make this book a good resource for young readers and some adults. For grades 7 and up.

Tagliaferro, L. *Genetic Engineering: Progress or Peril?* Minneapolis: Lerner Publishers, 1996. 128 pp. ISBN 0-822-52610-7.

The author balances the potential benefits of human, plant, and animal genetic experimentation with cautionary statements. Separate chapters on each of these topics follow a short introduction on the structure and functions of cells and DNA, and descriptions of genetics and the Mendelian theory of heredity. Discussion of human gene identification, the development of new life forms, and the regulation of genetic engineering leads into considerations of patent usage and how the exclusive use of genetic modifications of current organisms can turn into a monopoly. A substantial bibliography is provided including articles from periodicals, corporate scientific reports, and monographs. These are a rich source for debating current issues in genetic engineering. For grades 7–9.

Thro, E. *Genetic Engineering.* New York: Facts on File, 1993. 128 pp. ISBN 0-8160-2629-7.

By the use of charts and line drawings the author covers the mandatory topics of cells, DNA structure, chromosomes, and genes. She outlines some uses that have already been found (as of 1993) for genetic engineering and warns of upcoming ethical problems in eugenics, such as the patenting of new life forms. For grades 7–12.

Van Loon, B. *DNA, The Marvelous Molecule.* Norfolk, England: Tarquin Publications, 1990. 32 pp. ISBN 0-906-21275-8.

The helical structure of DNA and the topology of the many structures that inhabit the microscopic world of molecular biology are often inadequately portrayed on the flat page. The three-dimensional heavy paper models provided by Van Loon to be cut out and assembled solve this problem. The models include a DNA helix, a bacteriophage with a packaged DNA helix, nucleotide hydrogen bonding pairs, and a protein-folding model that graphically illustrates the effect of a mutation on a protein's shape. A 27-page minibook included within the package succinctly and clearly explains the models and the place of DNA in life, genetics, and evolution. This book is written for middle and senior high school students.

Wekesser, C., editor. *Genetic Engineering: Opposing Viewpoints.* San Diego: Greenhaven Press, 1997. 240 pp. ISBN 1-565-10358-0.

A discussion of the benefits and risks of the application of genetic engineering technology to medicine, agriculture, industry, and the environment. Ethical issues such as human cloning and genetic testing are also covered. For grades 5–12.

Wells, D. K. *Biotechnology.* Terrytown, NY: Bench Mark Books, 1997. 64 pp. ISBN 0-761-40046-X.

A description of the tools and wonders of biotechnology. For grades 3–5.

# U.S. Government Publications

## Office of Technology Assessment Documents

These documents are available from the U.S. Government Printing Office National Technical Information Service, Superintendent of Documents, 5285 Port Royal Road, Department 33, Springfield, VA 22161-0001.

*An Assessment of Alternatives for a National Computerized Criminal History System.* OTA-CIT-161, October, 1982.

*Biomedical Ethics and U.S. Public Policy. Hearing of the Committee on Labor and Human Resources.* OTA-BP-BBS-105, June, 1993.

*Biotechnology in a Global Economy.* OTA-BA-494, October, 1991.

*Commercial Biotechnology: An International Analysis.* OTA-BA-218, January, 1984.

*Cystic Fibrosis and DNA Tests: Implications of Carrier Screening.* OTA-BA-532, August, 1992.

*Field Testing Engineered Organisms: Genetic and Ecological Issues—Special Report.* OTA-BA-350, May, 1988.

*Genetic Counseling and Cystic Fibrosis Carrier Screening: Results of a Survey.* OTA-BP-BA-97, September, 1992.

*Genetic Counseling and Cystic Fibrosis Carrier Screening: Results of a Survey* (background paper). OTA-BP-BA-98, September, 1992.

*Genetic Monitoring and Screening in the Workplace: Contractor Documents.* OTA-BA-455, October, 1990.

*Genetic Screening in the Workplace.* OTA-BA-456, October, 1990.

*Genetic Witness: Forensic Uses of DNA Tests.* OTA-BA-483, July, 1990.

*Human Gene Therapy: A Background Paper.* OTA-BP-BA-32, December, 1984.

*Impacts of Applied Genetics: Micro-Organisms, Plants, and Animals.* OTA-HR-132, April 1981.

*Issues Relevant to NCIC (National Crime Information Center) 2000 Proposals.* Staff paper, November, 1987.

*Mapping Our Genes: Genome Projects—How Big? How Fast?* OTA-BA-373, April, 1988.

*Medical Testing and Health Insurance.* OTA-H-384, August, 1988.

*New Developments in Biotechnology: Public Perceptions of Biotechnology.* OTA-BP-BA-45, May 1987.

*Ownership of Human Tissues and Cells—Special Report.* OTA-BA-337, March 1987.

*Patenting Life—Special Report.* OTA-BA-370, April, 1989.

*Splicing Life: A Report on the Social and Ethical Issues of Genetic Engineering with Human Beings.* Pr 40.8: Et 3/L 62. November, 1982.

*Technologies for Detecting Heritable Mutations in Human Beings.* OTA-H-298, September, 1986.

*U.S. Investment in Biotechnology: An International Analysis.* OTA-BA-360, July, 1988.

# Congressional Reports

*Biotechnology Science Competitiveness Act of 1988.* U.S. House Committee on Agriculture. CIS 88 #703-15 (part 1); CIS 88 #163-18 (part 2).

*Biotechnology and the Ethics of Cloning: How Far Should We Go?* Hearing before the Committee on Science, Subcommittee on Technology, U.S. Congress, 1997. 59 pp. ISBN 0-160-55267-2.

*A Coordinated Framework for the Regulation of Biotechnology.* Fed. Reg. 51: 23303–23309, 1986.

*Designing Genetic Information Policy: The Need for an Independent Policy Review of the Ethical, Legal, and Social Implications of the Human Genome Project (16th report).* #102-478 USGPO. House Committee on Government Operation, U.S. Congress, 1992.

*The Genome Project: The Ethical Issues of Gene Patenting.* Hearing before the Committee on the Judiciary, Subcommittee on Patents, Copyrights, and Trademarks. 1993. 240 pp. ISBN 0-160-41610-8.

*Guidelines for Research Involving Recombinant DNA Molecules.* Fed. Reg. 59: 34496–34547, 1994.

*Issues in the Federal Regulation of Biotechnology: From Research to Release.* U.S. Congress House Committee on Science and Technology, Subcommittee on Investigations and Oversight. December, 1986. CIS 86 #702-18, 118 pp.

# Other Government Agency Reports

Special Projects Unit of the Bureau of National Affairs. *Biotechnology Law for the 1990s: Analysis and Perspective.* The BNA Special Report Series on Biotechnology, Special Report #4. Bureau of National Affairs, Washington, DC, 1989. 52 pp. ISBN 1-558-71153-8.

Blair, R. R. *Forensic DNA Analysis: Issues.* Washington, DC: U.S. Department of Justice, Office of Justice Programs, Bureau of Justice Statistics, 1991. 32 pp. GPO item # 0968-H-12 Doc. J 29.9/8:F 76.

*Recombinant DNA Research Volume 20: Documents relating to "NIH Guidelines for Research Involving Recombinant DNA Molecules," August 1994 to December 1994.* NIH Publication No. 95-3993, U.S. Department of Health and Human Services, Public Health Service, NIH Report of Recombinant Advisory Committee, December, 1995.

*Report on National Biotechnology Policy, The President's Council on Competitiveness.* Government document number Pr 41.8:B52 1991.

The President's Council on Competitiveness. *Report on National Biotechnology Policy.* Government document number Pr 41.8:B52, 1991.

*U.S. Biotechnology: A Legislative and Regulatory Roadmap, By the Special Projects Unit of the Bureau of National Affairs, The BNA Special Report Series on Biotechnology.* Special Report #2. Bureau of National Affairs, Washington, DC, 1989. ISBN 1-558-71139-2.

# Nongovernment Reports

*The Evaluation of Forensic DNA Evidence: An Update.* National Research Council, Committee on DNA Forensic Science. Washington, DC: National Academy Press, 1996. 254 pp. ISBN 0-309-05395-1.

*Genetic Information and Insuring: Confidentiality Concerns and Recommendations.* American Council of Life Insurance, Subcommittee on Privacy Legislation, 1990.

*Genetic Tests and Health Insurance: Results of a Survey.* Upland, PA: Dione Publishing Co., 1993. ISBN 1-56806-637-6.

Miller, T. H. *The Human Genome Project and Genetic Testing: Ethical Implications.* Series on The Genome, Ethics, and the Law: Issues in Genetic Testing. Washington, DC: American Association for the Advancement of Science, 1991.

*Report of the ACLI-HIAA Task Force on Genetic Testing.* Health Insurance Association of America, 1991.

# Periodicals and Newsletters

*The Ag Bioethics Forum.* 115 Morrill Hall, Iowa State University, Ames, IA 50011.

Interdisciplinary coverage of agricultural bioethics.

*Applied Genetic News.* Business Communications Company, Inc., 25 Van Zant Street, Norwalk, CT 06855-1781.

First published in 1980, this monthly newsletter is concerned with the application of genetic research to industry and technology. It evaluates ongoing research in cancer and other diseases and follows progress in the study of aging.

*BIO News.* Biotechnology Industry Organization, 1625 K Street NW, Suite 1100, Washington, DC 20006.

This newsletter concentrates on federal regulations and legislative developments affecting the biotechnology industry. Since 1985 it has served as the Association of Biotechnology Companies newsletter. Six issues per year, free to members.

*BioTech Market News & Strategies.* Conmar Enterprises, Inc., P.O. Box 11155, Ft. Lauderdale, FL 33339.

First published in 1983, this monthly newsletter contains marketing and product development strategies for biotechnology and pharmaceutical executives. In addition it covers "new marketable applications" and summarizes lawsuits related to biotechnology, patents, and environmental regulations.

*Biotech Reporter.* Freiberg Publishing, P.O. Box 7, Cedar Falls, IA 50613.

Formerly *AgBiotechnology News* and published since 1984, the *Biotech Reporter* is a monthly that covers business and technological aspects of international agricultural biotechnology. Besides the industrial perspective it also covers educational opportunities related to the field.

*Biotechnology and Development Monitor.* University of Amsterdam, The Netherlands: (joint) Directorate-General, International Cooperation of the Ministry of Foreign Affairs, The Hague, Switzerland and The University of Amsterdam, The Netherlands. Department of Political Science, Oudezijds Achterburgwal 237, 1012 DL Amsterdam. Website http://www.pscw.uva.nl/monitor.

International monitoring of biotechnology.

*Biotechnology Law Report.* Mary Ann Liebert, Inc., 20 W. 3rd Street, 2nd Floor, Media, PA 19063-2824.

This hefty bimonthly newsletter, with up to 200 pages per issue, covers legal developments in the fields of biotechnology and genetic engineering. First published in 1982, it includes aspects of product liability, patent, biomedical, contract and licensing, and international law along with pertinent legislation, regulatory actions, litigation resolution, and international developments. It publishes complete texts of significant court decisions, briefs, regulations, and legislation. Aimed primarily at lawyers with biotechnology clients, regulatory affairs professionals, and university and company biotechnology research departments.

*Biotechnology News.* CTB International Publishing, Inc., P. O. Box 218, Maplewood, NJ 07040-0218.

This newsletter concentrates on developments that impact genetic engineering, microbial and enzyme technology, and fermentation contributions to the production of pharmaceuticals, foods, crops, fuels, and chemicals. Established in 1980, this publication is mostly targeted at biotechnology industry management. Thirty issues per year, indexed semiannually.

*Biotechnology Newswatch.* The McGraw-Hill Companies, 1221 Avenue of the Americas, 36th Floor, New York, NY 10020.

This semimonthly newsletter is intended to provide an overview of the international biotechnology industry through a series of capsule summaries of news stories and a compilation of research and business stories. It is aimed at biotechnology industry professionals.

*BLAST: Bulletin of Law, Science & Technology.* Section of Science and Technology—American Bar Association, 750 N. Lake Shore Drive, Chicago, IL 60611.

This professional quarterly newsletter of the American Bar Association, established in 1976 as the *Bulletin of Law, Science & Tech-*

*nology,* is concerned with current issues relating law and science and technology. Subjects include the use of computers and the law, controls on scientific information, legal problems of genetic technologies, and policy issues in the communications industry.

**Carolina Tips.** Carolina Biological Supply Company, 2700 York Road, Burlington, NC 27215.

This quarterly newsletter contains articles of interest to science teachers from the elementary to college levels. Published by Carolina Biological Supply, a major source for science teaching supplies since 1938, it contains many useful tips for teaching, well illustrated articles, and student experiments. Indexed annually, it is free to science teachers (on written request on school letterhead) and to health professionals.

**DIALOG.** The Dialog Corporation. 11000 Regency Parkway, Suite 400, Cary, NC 27511. (Website: http://www.dialog.com)

This monthly newsletter covers the application of genetic research to industry and technology and evaluates ongoing research in aging, cancer, and other diseases. Research funding, venture capital, and stock prices are also discussed. First published in 1980, the targeted audience is both biotechnology professionals and laypersons.

**The Gene Exchange: A Public Voice on Genetic Engineering.** National Biotechnology Policy Center, National Wildlife Federation, 1400 16th Street NW, Washington, DC 20036.

Public forum on genetically modified organisms and the envirnoment.

**Genetic Engineering News.** Mary Ann Liebert, Inc., 2 Madison Avenue, Larchmont, NY 10538.

Current news on science and industry developments in genetic engineering technology. Includes business information, new companies, new products. Commentary and articles on issues of interest to the genetic engineering and biotechnology community.

**GeneWatch.** 19 Garden Street, Cambridge, MA 02138

Newsletter of the Council for Responsible Genetics, a resource for public involvement.

*Human Genome News.* Human Genome Management Information System for the U.S. Department of Energy, Oak Ridge National Laboratory, 1060 Commerce Park, Oak Ridge, TN 37830.

Formerly the *Human Genome Quarterly* and first published in 1989, this free governmental quarterly newsletter specializes in news about the Human Genome Project and is intended to facilitate communication among genome researchers. Besides the researchers, consumers of human genome information such as teachers, genetic counselors, physicians, ethicists, students, congressional staff, and the general public may find the information useful.

*Intellectual Property & Biodiversity News.* Institute for Agriculture and Trade Policy (IATP), 2105 1st Avenue South, Minneapolis, MN 55404

Aimed at biotechnologists, this free newsletter discusses the current news headlines that affect biotechnology.

*LEXIS Document Services.* 801 Adlai Stevenson Drive, Springfield, IL 62703-4261. (Website: http://www.netlds.com)

A comprehensive update since 1981 of public records of scientific, commercial, and governmental significance in the biotechnology field. Areas included are genetic engineering, hybridoma technology, applied plant genetics, enzymology, and biomass conversion. It is aimed at a wide audience interested in the progress and future of biotechnology, from scientists to investment analysts, patent attorneys, and government officials.

*Life Sciences & Biotechnology Update.* Infoteam, Inc., P.O. Box 15640, Plantation, FL 33318-5640.

Aimed at professionals involved with the applications of biotechnology to the life sciences, this monthly newsletter covers activities in agriculture, food, medicine, health, biological and biomedical engineering, microorganisms, nutrition, and disease.

*McGraw-Hill's Biotechnology Newswatch.* McGraw-Hill, Inc., 1221 Avenue of the Americas, 36th Floor, New York, NY 10020.

Features news of developments in science, finance, commerce, and government relevant to genetic engineering, applied plant genetics, industrial process biotechnology, and biomass conversion. Published semiannually since 1981.

*Mealey's Litigation Report: Biotechnology.* Mealey Publications, Inc., 512 W. Lancaster Avenue, P. O Box 446, Wayne, PA 19087-0446.

One of several newsletters specializing in different topics of legal litigation, the semimonthly *Report* beginning in 1996 has covered in depth the legal disputes over processes and products developed through the use of biotechnology "from DNA to drugs to disease detection." Legislative action and international disputes as well as expert commentary are included. It is directed primarily to legal counsel for biotechnology, chemical, and pharmaceutical companies.

*New England Regional Genetics Group—Regional Newsletter.* New England Regional Genetics Group, P.O. Box 670, Mt. Desert, ME 04660.

First published in 1985, this free semiannual newsletter covers news of interest to genetic service professionals and consumers of genetic services in the New England region.

*Olsen's Biotechnology Report.* G. V. Olsen Associates, 123 Picketts Ridge Road, West Redding, CT 06896.

This monthly report is focused primarily on food and agricultural biotechnology. It provides coverage on genetic engineering research, scaleup, production, and marketing of plant, animal, and agricultural chemicals. It also includes a financial analysis of the biotechnology market and investment opportunities in that area. First published in 1983, this newsletter lists large farmers and growers as well as universities as a part of its audience.

*U.S. Regulatory Reporter.* Parexel International Corporation, 195 West Street, Waltham, MA 02154-1116.

This monthly newsletter, first published in 1984, covers regulatory news from the Food and Drug Administration, primarily for pharmaceutical and biotechnology industry professionals. It provides analyses of FDA product approval standards and the review process.

*West Coast Biowatch.* 2700 H Street, Sacramento, CA 95816.

*Your World: Biotechnology and You.* Pennsylvania Biotechnology Association, 1524 W. College Avenue, Suite 206, State College, PA 16801.

This journal for junior and senior high students describes the application of biotechnology, science, and engineering to problems facing our world.

# Directories

Alford, II, J. E., editor. *Genetic Engineering and Biotechnology Firms Worldwide Directory.* Princeton Junction, NJ: Mega-Type Publishing. 703 pp. $299. (Also available in alternate format: diskette with custom search and report software, $599.99.)

International coverage of 6,000 firms with biotech divisions as well as small, independent companies. This book is also available on diskette. Information includes: company name, address, division name (if applicable), name and title of executives, research lab locations, number of employees engaged in biotech or genetic engineering, equity interests held by others, areas of research activity, and currently available products. *Annual, October.*

Barry, Inc. *National Biotech Register.* Wilmington, MA: Barry, Inc., 1995. 242 pp. $58.

Coverage of 2,200 companies active in biotech research, development, and manufacturing. This book is also available on microfiche. Information includes: company name, address, phone, fax, names and titles of key personnel, description of business activities, number of employees, year founded.

Brogna, C., editor. *Bioindustry Directory.* Maplewood, NJ: CTB International Publishing Co., Inc. 165 pp. $157/year.

A listing of more than 3,000 U.S. and foreign companies, government agencies, cultural collections, professional organizations, and trade groups working in genetic engineering, plant biotechnology, applied molecular biology, and bioculture technologies. Information includes: name of firm or agency, address, phone, fax. *Annual.*

Coombs, J., and Y. R. Alston, editors. *Biotechnology Directory.* New York: Stockton Press. 500 pp. $275, plus shipping and handling.

Lists more than 10,000 companies, universities, research centers, government agencies, and suppliers of products and services.

Information includes: organization name, address, phone, fax, contact, and description of products, services, or research. *Annual, December.*

Crafts-Lighty, A., E. Burak Reed, and J. Sime. *UK Biotechnology Handbook.* Slough, Berkshire, England: BioCommerce Data Ltd. 620 pp. $180 + $30 airmail shipping. Also available through online vendor: DIALOG (Knight-Ridder Information, Inc.), as *BioCommerce Abstracts and Directory DataStar.*

Covers more than 700 biotech companies, research institutes, universities, and related professional and academic associations. This book is also available on diskette and as mailing labels. Information includes: organization name, phone, telex, fax, names and titles of key personnel, number of employees, description of activities, and areas of research interest. *Every 18 months.*

Dibner, M. D., editor. *Biotechnology Guide, Third Edition.* Research Triangle Park, NC: Institute for Biotechnology Information, 1994. 692 pp.

This book is also available as an electronic database. The U.S. Companies Database includes 1,100 companies; the Actions Database includes more than 10,000 strategic actions and alliances in commercial biotechnology from 1981; other databases are on general topics of interest to commercial biotechnology. Information includes: address, phone, fax, financing, names and titles of key management personnel, products on market and in R&D, number of employees, technologies used, partnerships, and failed companies. *Irregular publication frequency; second edition in 1991, third edition in 1994.*

Han Consultants. *Chinese Biotechnology Directory.* Wuhan, Hubei, People's Republic of China: Han Consultants, 1993. 256 pp. ISBN 7-5011-0570-9.

Biotechnology in the People's Republic of China. This directory claims to be the only collected source in English on biotechnology in mainland China. Information includes: lists of companies, location, products, areas of interest, and contacts. It also provides an overview of biotechnology in mainland China, a description of government policy on technology transfer, intellectual property and patent coverage, and a list of information resources in China—abstracting services, journals, and newsletters.

McGrath, K. A., editor. *Who's Who in Technology.* Detroit, MI: Gale Research, 1701 pp. $195.

This directory lists more than 25,000 North American men and women working in more than 1,000 scientific and technology fields. Also available as an online database through LEXIS-NEXIS or GALBIO (vendor: ORBIT Search Service, WHOTECH). Information includes: name, occupation, personal data, educational background, career information, technical achievements, organizational affiliations, honors, awards, special achievements, area(s) of expertise, technical publications, patent data, and addresses. *Irregular.*

Ratafia, M., editor. *TMG's Worldwide Biotechnology and Pharmaceutical Desk Reference.* New Haven, CT : Technology Management Group, 1311 pp. $372 (print edition). Also available in alternate formats (diskette $338, or $398 with print edition included).

The listing covers more than 375 biotech and pharmaceutical companies, 321 universities, 128 U.S. government agencies and labs, and 402 company funding sources. This book is also available on diskette. Information includes: company/institution name, address, phone, fax, and name and title of contact. *Annual.*

Self, J. *North American Biotechnology Directory.* Houston: I.E.I. Publishing Division. 620 p. $89.

This Directory lists companies, research institutions, universities, government agencies, suppliers, and manufacturers involved with biotechnology throughout the U.S., Canada, and Mexico. This book is also available on diskette and as mailing labels. Information includes: company/institution name, address, contacts, phone, fax. *Annual.*

U.S. Department of Commerce Technology Administration. *Directory of Federal Laboratory and Technology Resources: A Guide to Services, Facilities, and Expertise.* Washington, DC: U.S. Department of Commerce Technology Administration, 1993. ISBN 0-93421-340-2.

A listing of federally funded programs provided by National Technical Information Service and National Technology Transfer Center and arranged by subject area. Information includes: organization name, institution, address, phone, name and title of contact, facilities or activities, and fields of emphasis.

Woolum, J., editor. *AgriBioScan.* Phoenix, AZ: Oryx Press. 320 pp. $395/yr.

A compilation of statistics on agricultural biotechnology companies. Information includes: company name, address, phone, fax, research and development, key personnel, financial data, partnerships, and products awaiting FDA or USDA approval. *Three times/year.*

# Selected Nonprint Resources

# 7

This chapter lists genetics, biotechnology, and genetic engineering resources available outside of the printed page. Educational videocassettes are reviewed. Some of them are designed as instructional aids; others are public lectures by leaders in the field or had a first life on educational television. Instructional slide sets on genetic engineering are also available.

Electronic handling of information has expanded with the increasing availability of powerful personal computers and the Internet. Computers allow for more varied presentation of material and the ability to search for specific information. Examples of computer educational software and CD-ROM contributions enhanced with pictures, text, and audio are provided. Also included are electronic databases and directories of information, sources on aspects of biotechnology and genetic engineering that are becoming increasingly popular because they are readily expanded and updated, and are easy to use with powerful search functions. The World Wide Web computer network, accessible from home computers and from computers in many schools and public libraries, is a fertile source for all types of information relevant to genetic engineering, ranging from the latest technical nuances to government documents and to personal opinions on issues. Meaningful use of this information requires that the consumer consider the source carefully as standards for verification of posted facts differ among the websites. Only a sampling of websites on general biotechnology is provided in this chapter as individual web addresses come and go. Searching for keywords such as *biotechnology, genetic,* or *recombinant DNA* will quickly generate a plethora of interesting sites.

# Videocassettes

### The Biological Revolution: 100 Years of Science at Cold Spring Harbor

| | |
|---|---|
| *Type:* | VHS |
| *Age level:* | Junior and senior high school |
| *Length:* | 30 minutes |
| *Cost:* | $19.95 |
| *Date:* | 1986 |
| *Source:* | Cold Spring Harbor Press |
| | 10 Skyline Drive |
| | Plainview, NY 11803-2500 |
| | (800) 843-4388 |
| | Fax: (516) 349-1946 |
| | E-mail: cshpress@cshl.org |
| | Website: http://www.cshl.org/ |

This video documentary of the quest to understand the nature of the genetic code also traces the parallel evolution of American science and society. James Watson, director of the Cold Spring Harbor Laboratories and a Nobel laureate for his elucidation of the structure of DNA, is joined by fellow laureates Alfred Hershey, Salvador Luria, Barbara McClintock, and Walter Gilbert to provide their own firsthand insight of the genetic revolution. Produced for the average nonscientist adult, it is suitable as supplementary material for junior and senior high school science classes. (ISBN 0-879-69991-4)

### Francis Crick: Beyond the Double Helix

| | |
|---|---|
| *Type:* | VHS |
| *Age level:* | High school through college |
| *Length:* | 28 minutes |

| | |
|---|---|
| *Cost:* | $49.95 |
| *Date:* | 1989 |
| *Source:* | Carolina Biological Supply |
| | 2700 York Road |
| | Burlington, NC 27215 |
| | (800) 334-5551 |
| | Fax: (800) 222-7112 |
| | Website: http://www.carosci.com/ |

Cowinner, with James Watson and physicist Maurice Wilkins, of the 1962 Nobel Prize for the double helical structure of DNA, Francis Crick describes his early interest in science and the events leading to the elucidation of the double helix. He recounts the key steps in the deduction process and his partnership with Watson. Other topics include his current scientific passions, the origin of life on earth, and the functioning of the conscious and unconscious brain. (Catalog #F6-53-6400-V)

**Genetic Testing**

| | |
|---|---|
| *Type:* | VHS |
| *Age level:* | Grades 6 through college |
| *Length:* | 26 minutes |
| *Cost:* | $149.00 |
| *Date:* | 1990 |
| *Source:* | Carolina Biological Supply |
| | 2700 York Road |
| | Burlington, NC 27215 |
| | (800) 334-5551 |
| | Fax: (800) 222-7112 |
| | Website: http://www.carosci.com/ |

This videotape demonstrates amniocentesis and fetal monitoring by ultrasound and discusses the advantages as well as the disadvantages of prior genetic knowledge about potential medical problems. (Catalog #F6-49-2228A)

**HuGEM (Human Genome Education Model)**
Tape 1: An Overview of the Human Genome Project and its
Ethical, Legal, and Social Issues
Tape 2: Opportunities and Challenges of the Human Genome
Project (Francis Collins, narrator)
Tape 3: Issues of Genetic Privacy and Discrimination
Tape 4: Genetic Testing across the Life Span
Tape 5: Working Together to Improve Genetic Services

*Type:*      VHS
*Age level:*   Unspecified (general public)
*Length:*   Set of five videotapes ranging from 19 to 45 minutes each
*Cost:*   $50.00 for the set or $15.00 each; includes a manual.
*Date:*   1997
*Sources:*   Georgetown University Child Development Center
3307 M Street NW, Suite 401
Washington, DC 20007-3935
E-mail: Laphamy@medlib.georgetown.edu
*or*
Alliance of Genetic Support Groups
4301 Connecticut Avenue NW, Suite 404
Washington, DC 20008
(202) 966-5557; (800) 336-GENE(4363)
Fax: (202) 966-8553
E-mail: info@geneticalliance.org

This set of videos is designed to educate the general public. It consists of a series of interviews with researchers, physicians, and consumers.

**The Life Revolution Series: On the Frontier of Biogenetics**
*Type:*      VHS
*Age level:*   Grades 6 through college
*Length:*   12 tapes, 26 minutes each
*Cost:*   $995/set; $89.95 each
*Date:*   1990
*Source:*   Carolina Biological Supply
2700 York Road
Burlington, NC 27215
(800) 334-5551
Fax: (800) 222-7112
Website: http://www.carosci.com/

This series of 12 videotapes, each 26 minutes long, describes the science of genetic engineering along with its promises and risks. Titles are: *Cutting and Splicing DNA; Evolution: Man Takes a Hand; The Human Genome; DNA Techniques; Designer Plants; Depleting the Gene Bank; Sowing the Seeds of Disaster; Growing Synthetics; Cell Wars; Recombinant Technology; Superanimals, Superhumans?* and *Whither Biogenetics?* (Catalog # F6-49-2228)

**Map of Life: Science, Society, and the Human Genome Project**
*Type:*        VHS
*Age level:*   High school through college
*Length:*      46 minutes
*Cost:*        $49.95
*Date:*        1992
*Source:*      Carolina Biological Supply
               2700 York Road
               Burlington, NC 27215
               (800) 334-5551
               Fax: (800) 222-7112
               Website: http://www.carosci.com/

Dr. James D. Watson, the former head of the National Center for Human Genome Research, explains the goals of the 15-year project to determine the nucleic acid sequence of the entire human genome. Molecular biologist Dr. Walter Gilbert explains some of the science behind the mapping and sequencing of genes and how genes control function in the body. The implications of the results of this project on ethical and legal issues and the impact of the knowledge base on medicine and biology are discussed by Dr. Louis Sullivan, secretary of Health and Human Services, and Dr. Bernadine Healy, Director of NIH. (Catalog # F6-49-2215-V)

**Molecular Miracles: Human Gene Therapy
and the Future of Modern Medicine**
*Type:*        VHS
*Age level:*   High school through college
*Length:*      47 minutes
*Cost:*        $49.95
*Date:*        1993
*Source:*      Carolina Biological Supply
               2700 York Road
               Burlington, NC 27215
               (800) 334-5551
               Fax: (800) 222-7112
               Website: http://www.carosci.com/

Leading gene therapy researchers, including Drs. W. French Anderson, Steven Rosenberg, Michael Blaese, Kenneth Culver, Ronald Crystal, Edward Oldfield, Zvi Ram, David Curiel, Louise Markert, Dusty Miller, and Nelson Wivel, describe the early attempts at gene therapy in the early 1990s, newer techniques employing retroviral vectors for gene insertion, and the outlook for this specialized therapy. Issues of government regulation and the legal and ethical considerations of this new form of medicine are also discussed. (Catalog # F6-53-6550-V)

**On Becoming a Scientist**
*Type:*        VHS with teachers' guide
*Age level:*   Young audience (unspecified age)
*Length:*      19 minutes
*Cost:*        $70.00
*Date:*        1996
*Source:*      Cold Spring Harbor Press
               10 Skyline Drive
               Plainview, NY 11803-2500
               (800) 843-4388
               Fax: (516) 349-1946
               E-mail: cshpress@cshl.org
               Website: http://www.cshl.org/

This video uses a series of interviews to portray a day in the life of three graduate students and a laboratory manager. The purpose is to dispel stereotypes and to provide role models. (ISBN 0-879-69486-6)

**Promise and Perils of Biotechnology: Genetic Testing**
*Type:*        VHS with teachers' guide
*Age level:*   High school through college
*Length:*      25 minutes
*Cost:*        $70.00
*Date:*        1996
*Source:*      Cold Spring Harbor Press
               10 Skyline Drive
               Plainview, NY 11803-2500
               (800) 843-4388
               Fax: (516) 349-1946
               E-mail: cshpress@cshl.org
               Website: http://www.cshl.org/

This video is a documentary of families at risk for the genetic diseases Huntington's disease and familial hypercholesterolemia. It provides a window into the reality of the conditions and first-hand exposure to the ethical, legal, and social dilemmas faced by families at risk for genetic disorders. Background information is provided in the printed guidebook that provides information on genetic disorders, related education/support organizations, student activities, a glossary, and a reference list. (ISBN 0-879-69493-9)

**Stories from the Scientists**

| | |
|---|---|
| *Type:* | VHS and teachers' guide |
| *Age level:* | High school and college, and the public |
| *Length:* | 30 minutes |
| *Cost:* | $50 |
| *Date:* | 1994 |
| *Source:* | University of California, San Francisco |
| | San Francisco, CA 94143 |

This video is a documentary that tells the story of two instrumental partnerships related to genetic engineering. Through animation, interviews, re-enactment, and documentary footage, it displays the scientific achievements, personalities, and approaches of the James Watson–Francis Crick team (structure of the DNA helix) and the Herbert Boyer–Stanley Cohen team (recombining DNA molecules and cloning genes). The 32-page Teacher's Guide expands on the tape and provides additional resources and activities for the classroom. (ISBN 0-879-69462-9)

**Winding Your Way through DNA**
Tapes 1 & 2: Discovering the Wonder of DNA
Tapes 3 & 4: New Ways to Use DNA
Tapes 5 & 6: Asking the Tough Questions about DNA Technology

| | |
|---|---|
| *Type:* | VHS |
| *Age level:* | High school and college |
| *Length:* | 8½ hours (6 tapes) |
| *Cost:* | $250 academic; $400 nonacademic |
| *Date:* | 1993 |
| *Source:* | University of California, San Francisco |
| | San Francisco, CA 94143 |

This is a documentary video set of a symposium on DNA technology held in September 1992. During the symposium leaders in the field addressed three questions: What is gene technology? How can it be used in industry and medicine? Does it have risks and dangers? A Q&A session with science journalists that took place after the symposium is included. References for additional reading are provided in the guidebook. (ISBN 0-879-69393-2)

# Slide Sets

**Elementary Genetics**

| | |
|---|---|
| *Type:* | 110 35-mm slides with a 27-minute narrative cassette and teachers' guide |
| *Age level:* | Grades 6 through college |
| *Cost:* | $247.50 |
| *Source:* | Carolina Biological Supply<br>2700 York Road<br>Burlington, NC 27215<br>(800) 334-5551<br>Fax: (800) 222-7112<br>Website: http://www.carosci.com/ |

These slides portray early discoveries of modern genetics following the work of Gregor Mendel. They offer a description of simple Mendelian genetics and conclude with a discussion of probabilities, the key to understanding population genetics. (Catalog # F6-48-1210)

**Genetic Engineering Set**

| | |
|---|---|
| *Type:* | 63 35-mm slides with a narrative cassette and teachers' guide (Set A); 20 35-mm slides and a printed guide (Set B) |
| *Age level:* | High school through college |
| *Cost:* | $141.75 (Set A); $45.00 (Set B) |
| *Source:* | Carolina Biological Supply<br>2700 York Road<br>Burlington, NC 27215<br>(800) 334-5551<br>Fax: (800) 222-7112<br>Website: http://www.carosci.com/ |

(Catalog # F6-48-1217A [Set A]; F6-48-1217B [Set B])

**Heredity I: Traits and Alleles Set**

| | |
|---|---|
| *Type:* | 60 35-mm slides with a narrative cassette and teachers' guide |
| *Age level:* | High school through college |
| *Cost:* | $135.00 |
| *Source:* | Carolina Biological Supply<br>2700 York Road<br>Burlington, NC 27215 |

(800) 334-5551
Fax: (800) 222-7112
Website: http://www.carosci.com/

Presents basic Mendelian genetics involving the concepts of genes, alleles, dominance, and recessiveness. (Catalog # F6-48-1211)

### Heredity II: Multiple Genes and Chromosomes
*Type:*          79 35-mm slides with a narrative cassette
                 and teachers' guide
*Age level:*     High school through college
*Cost:*          $177.75
*Source:*        Carolina Biological Supply
                 2700 York Road
                 Burlington, NC 27215
                 (800) 334-5551
                 Fax: (800) 222-7112
                 Website: http://www.carosci.com/

This set of slides covers more advanced genetic concepts. Two- and three-gene interactions are discussed, including genetic linkage and genetic mapping. These population genetics tools were precursors of the modern Human Genome Project. They illustrate the power of logical approaches in deriving a wealth of information before genes and DNA were understood at the molecular level.

# Computer Programs and Other Electronic Resources

### DNAdetective
*Type:*          CD-ROM
*Computer:*      IBM or compatible 386, Windows 3.1 or higher, 4
                 MB RAM, SVGA monitor, CD-ROM drive
*Age level:*     Grades 9 through college
*Cost:*          $99.00
*Source:*        Carolina Biological Supply
                 2700 York Road
                 Burlington, NC 27215
                 (800) 334-5551
                 Fax: (800) 222-7112
                 Website: http://www.carosci.com/

This program teaches students about DNA technology that is used in forensic science to provide legal evidence. It simulates such topics as DNA extraction, polymerase chain reaction, restriction enzyme digests, DNA electrophoresis, fragment analysis, and others. The results are then interpreted as they would be in a court of law and the students judge the validity of the evidence. The program can be used at either beginner or DNAdetective skill levels. A comprehensive teachers' guide is provided. (Catalog # W1-39-9158)

**Exploring Genetics and Heredity**
*Type:*        CD-ROM
*Computer:*   Macintosh with color monitor, 2 MB RAM,
              System 6 or 7, CD-ROM drive; or IBM-compatible
              MS-DOS, MCGA or VGA monitor, Soundblaster
              or compatible sound card, CD-ROM drive
*Age level:*  Grades 8 and up
*Cost:*       $100.00
*Source:*     Carolina Biological Supply
              2700 York Road
              Burlington, NC 27215
              (800) 334-5551
              Fax: (800) 222-7112
              Website: http://www.carosci.com/

This interactive multimedia CD-ROM describes DNA structure, cellular replication, and the composition and regulation of genetic material. Classical Mendelian heredity is also covered. Includes a teachers' guide. (Catalog # F6-39-9080 [Mac]; F6-39-9081 [MS-DOS])

**The Nature of Genes**
*Type:*        CD-ROM
*Computer:*   Macintosh, 4 MB RAM, System 6.0.7 or higher,
              CD-ROM
*Age level:*  High school through college
*Cost:*       $175.00
*Source:*     Silver Platter Education
              100 Ridge Drive
              Norwood, MA 02062-5043
              (781) 769-2599
              Fax: (781) 769-8763

This CD-ROM, designed to be user-interactive, is both a general genetic education tool and a reference tool. The tutorial covers

DNA, DNA replication, the genetic message, the genetic code, genes, transcription, translation, proteins, inheritance, and the Human Genome Project. (ISBN 1-57276-001-X; order # SE-001-001)

**Occupational Outlook Handbook**
*Type:*       CD-ROM
*Computer:*   Macintosh or Windows PC; CD-ROM drive
*Age level:*  High school
*Cost:*       $399.00
*Source:*     Carolina Biological Supply
              2700 York Road
              Burlington, NC 27215
              (800) 334-5551
              Fax: (800) 222-7112
              Website: http://www.carosci.com/

After an interactive session in which the software surveys the student's educational interests and personality traits, the program provides information from over 300 career titles on necessary skills, high school courses, length of training, common outlook, related work activities, related occupations, and where to go for additional information. The CD-ROM format includes short video clips for groups of careers, giving the students a visual feel for the work environment. Regular upgrades of new career fields are planned. (Catalog # W1-39-4996)

**Women in Science**
*Type:*       CD-ROM
*Computer:*   Macintosh or Windows PC; CD-ROM drive
*Age level:*  Grades 5 through 12
*Cost:*       $79.95
*Source:*     Carolina Biological Supply
              2700 York Road
              Burlington, NC 27215
              (800) 334-5551
              Fax: (800) 222-7112
              Website: http://www.carosci.com/

The advantages of the interactive CD-ROM format are evident in the compelling stories of current women scientists and their work. Twelve interview questions are available to hear about these women's lives and their love for science as the students "visit" them where they work. Interactive experiments allow the student to join scientific teams collecting data and analyzing results. A database of information on 130 past and present women scientists is also included. (Item #W1-39-83316)

# Databases

### BioCommerce Abstracts and Directory
Crafts-Lighty, A., editor. Slough, Berkshire, England: BioCommerce Data, Ltd. The online vendor is DIALOG (Knight-Ridder Information, Inc. Vendor: DataStar (Knight-Ridder Information, Ltd., 18/20 Hill Rise, Richmond, Surrey TW10 6UA England; biocom@dial.pipex.com). Alternate format on diskette. Semimonthly.

This database contains 135,000 abstract records on 2,000 U.S. and European biotech companies, research institutes, universities, and related professional and academic associations. Information includes: name, occupation, personal data, educational background, career information, technical achievements, organizational affiliations, honors, awards, special achievements, area(s) of expertise, technical publications, patent data, addresses, production/services, and research areas.

### Directory of Biotechnology Information Resources
Specialized Information Services Division
U.S. National Library of Medicine
8600 Rockville Pike
Bethesda, Maryland 20894
http://www.tdnettolocater.nim.nih.gov
Quarterly

This online database contains more than 3,300 records with information on computer bulletin boards and networks, culture collections, specimen banks, biotechnology centers and related organizations, periodicals, directories, monographs, nomenclature reconciliation, and assorted sources of biotechnology information. Information includes: title, organization name, address, phone, name and title of contact, facilities or activities, fields of emphasis, language, limitations of use/availability of resource.

### WooBioScan: The Worldwide Biotech Industry Reporting Database
Woolum, J., editor. Phoenix, Arizona: Oryx Press. Online vendor is Knowledge Express data systems (http://www.knowledge express.com). Alternate formats: mailing labels, magnetic media, complete database, mailing list. Updated monthly.

The database contains more than 1,000 companies doing product research and development in food processing, agriculture, medicine, and other fields in biotechnology. Information includes: company name, address, phone, names and titles of key personnel, number of employees (including number of Ph.D.s), date founded, names and descriptions of subsidiaries, names of investors and percentage of investment, names and descriptions of agreements and contracts. Annual, with bimonthly supplements. $975/yr including supplements.

# Environmental Remediation Databases

### Alternative Treatment Technology Information Center (ATTIC)
System operator: (301) 670-6294
Online access: (301) 670-3808

This EPA database is concerned with environmental treatment technologies. Information includes: treatment technologies, treatability of different contaminants, technical assistance sources, and a calendar of conferences. An electronic bulletin board is available.

### Cleanup Information Bulletin Board (CLU-IN)
System operator: (301) 589-8368
Online access: (301) 589-8366

This bulletin board of the Office of Solid Waste and Emergency Response is concerned with remediation of Superfund and RCRA corrective action sites.

### Computerized On-Line Information Service (COLIS)
System operator: (908) 589-8368
Online access: (908) 548-4636 (password: EPA)

This online system provided by EPA offers a database and free information on toxic waste site cleanups. It includes information on case histories, library search systems, site application analysis reports, and Risk Reduction Engineering Laboratory (RREL) treatability.

### Vendor Information System for Innovative Treatment Technologies (VISITT)
$5^1/_4$" or $3^1/_2$" floppy disk, for DOS 3.3 or higher. For copies call (800) 245-4505 or (703) 883-8448.

This system provides information on various commercially available technologies for bioremediation, chemical and thermal treatment methods. Information provided by vendors is not verified or endorsed by the EPA. The EPA database is free.

# Internet Sources on Genetic Engineering

These addresses are meant to provide an entrée to the wealth of information accessible on the Internet. Interconnections with related sites are easily explored by clicking on the built-in links. The search function on your Web browser will quickly locate many more.

**Access Excellence**
Genentech, Inc., San Francisco, California.
http://www.gene.com/ae

A national educational program sponsored by the biotechnology company Genentech and described in the site header as "A Place in Cyberspace for Biology, Teaching, and Learning." The site provides a variety of information, including science updates, clips on the researchers, forensic scientists, and other people contributing to the ongoing discussion of bioethics, the history of recombinant DNA, and other issues. A variety of activities for teachers and their students are suggested.

**Agbiotech Online**
Information Systems for Biotechnology, Virginia Tech
http://gophisb.biochem.vt.edu/

Information on agricultural and environmental biotechnology research, product development, regulatory issues, and biosafety.

**Bioethics Journals, Networks, and Associations
in the United States and Canada**
The Center for Bioethics, University of Pennsylvania, Philadelphia
http://www.med.upenn.edu/~bioethic/outreach/bioforbegin/organizations.html

Lists of organizations and publications concerned with bioethics, and links to other sites.

**Bio Online**sm
Vitadata Corporation
http://www.bio.com/

A comprehensive site for information and services related to biotechnology. It includes resources on industry, government, nonprofit special interest groups, research, career information, and education.

**BioTechnology Permits Home Page**
United States Department of Agriculture
http://www.aphis.usda.gov/BBEP/BP/

Lists of links to other biotechnology websites, both U.S. government and international, at /links.html. A good source for information on regulations pertaining to agricultural biotechnology. Provides access to a database (1987–present) on applications for release or testing of bioengineered crops.

**Blazing a Genetic Trail**
Howard Hughes Medical Institute
http://www.hhmi.org/GeneticTrail/

This nontechnical educational site sponsored by the Howard Hughes Medical Institute is designed to educate the general public. It is organized as a series of well-written, illustrated stories such as "How Genetic Disorders are Inherited" and "Stalking a Lethal Gene."

**The Gene Letter**
The Shriver Center
http://www.geneletter.org/mainmenu.htm

An all-inclusive site for information on the Web about scientific and social issues in genetics. It has links to federal and state genetics agencies and legal statutes. It includes an uncensored chat area and a search engine for locating information elsewhere in the site and on the Web.

**The Pure Food Campaign**
Foundation of Economic Trends
http://interactivism.com/purefood/

Website for an activist group focusing on issues of food production—organic vs. genetically engineered or processed. The Pure

Food Campaign is allied with Jeremy Rifkin's Foundation on Economic Trends. Links to other similar types of groups can be found at http://interactivism.com.

**United States Department of Agriculture**
http://www.ice.net/jumps/ag/ag.html

Includes many links to agricultural biotechnology sites.

# Glossary

## Definitions

**aerobe** An organism that can grow only in the presence of oxygen.

**amino acid** A chemical building block that is used by cells to make proteins and is converted into other chemicals needed by the cell. There are twenty different natural amino acids in proteins.

**anaerobe** An organism that cannot grow in the presence of oxygen.

**antibiotic** A chemical produced by one organism that kills or inhibits growth of another organism in competition for food or space. Often produced by microorganisms (yeast or bacteria). Humans have adopted the use of antibiotics to control organisms that cause disease.

**antibodies** Defense proteins produced by the body in response to either vaccines or the organism or a component. They bind and neutralize the toxic agent, allowing the body to clear the danger.

**aquifer** A natural underground collection of water accumulated in a layer of permeable rock or sand from percolating surface water usually cleansed by its passage through soil layers. Polluted soils are leading to contamination of the underground water, which normally has a very slow turnover. This pool is often the source of fresh water for human consumption or for agriculture.

**aromatic** An organic chemical compound with a very low hydrogen-to-carbon ratio. Because of their resistance to biological modification aromatic compounds can be quite stable in the environment and are often considered pollutants. Examples are benzene, phenol, naphthalene, dioxin, 2,4-D, and biphenyls.

**bacteria (bacterium,** singular) Single cell microorganisms, capable of multiplying independently, whose genetic information is not enclosed in a membrane-bounded nucleus. Bacteria are thought to be one of the oldest forms of life.

**bacteriophage** (also **phage**) A virus that infects bacteria. Used by molecular biologists as a vehicle to transfer DNA sequences into host bacteria for study.

**base pair** Technical term that refers to the nucleotides paired across the double-stranded DNA helix by hydrogen bonds (A=T; C≡G). One nucleotide in a single DNA (or RNA) sequence corresponds to one base pair (CAT or CG) in a double-stranded DNA (or RNA) sequence.

**biodegradable** Capable of being broken down into its component parts by biological organisms.

**bioethics** Application of the ideas of ethical conduct to biotechnology and medicine. Trying to reconcile the ability to accomplish certain medical or genetic or environmental actions with the societal norms of acceptable actions. Being able to perform a genetic test doesn't mean that it should be done, especially if there is no cure for the disease. Some people consider doing a genetic test for an incurable disease once a person is born to be unethical.

**biomass** Total dry weight of biological organisms.

**bioremediation** The use of biological organisms to remove toxins and wastes from contaminated materials.

**catalyst** An agent that speeds up the rate of a chemical reaction. In cells catalysts are special proteins called enzymes that allow reactions to occur at body temperature.

**central dogma** A hypothesis stating that the direction of genetic information flow is from DNA through RNA (messenger RNA) into protein. This turns out to hold for most situations except, notably, for certain viruses that contain RNA as their genetic material. These are called retroviruses, referring to their ability to replicate their RNA through a DNA intermediate that they generate by the action of a reverse transcriptase enzyme.

**chromosomes** Structures in the nuclei of cells that consist of DNA wrapped for compactness on a protein scaffold. The DNA in chromosomes contains the linear sequence of genes. Thus, a gene resides in a particular position on a particular chromosome in all cells of an individual in a species. Human cells contain twenty-three pairs of chromosomes. Reproductive cells contain only one copy of each chromosome so that when the sperm and ovum combine, the resulting zygotic cell (soon to become an embryo) contains one set of chromosomes (and genes) from each parent.

**clone**   A genetic replica, either an organism or a piece of DNA.

**codon**   A group of three consecutive nucleotides in the DNA polymer (sixty-four possibilities) that codes either for one of the twenty amino acids or peptide chain termination.

**cytoplasm**   The semisolid substance that fills plant, animal, and microbial cells and in which are immersed the nucleus material and other cellular structures that carry out the chemistry of life.

**deoxyribonucleic acid (DNA)**   A chain or polymer that contains the genetic information for an organism. It is composed of four different kinds of units called nucleotides: adenosine (A), thymidine (T), cytosine (C), and guanosine (G).

**DNA ligase**   An enzyme that chemically joins separate (usually restriction enzyme-cleaved) double-stranded DNA segments together end-to-end to produce a continuous DNA molecule. An important tool of recombinant DNA technology.

**dominant gene**   A genetic characteristic that is expressed when present in either of the set of chromosomes from the parents carrying that DNA sequence. In genes controlling eye color, brown eyes are a dominant characteristic.

**dominant trait**   A single-gene characteristic that is expressed when the gene is present on only one of the two chromosomes carrying the particular gene. Among genes controlling eye color, the gene for brown eyes is dominant.

**ecology**   The study of the interaction of groups of organisms with each other and with their environment.

**ecosystem**   A unit consisting of a natural community of organisms together with their environment.

**enzyme**   A protein that speeds up, or catalyzes, a chemical reaction and controls the production of the product(s) of the reaction.

**ethics**   A philosophy or system for judging right and wrong actions in human conduct. Sometimes termed *morality*, the basic assumptions about what constitutes right and wrong are often difficult to agree upon within any particular society. Complicating the issue is that these basic assumptions can vary dramatically between societies with different histories. Since they are usually deeply held convictions rather than verifiable facts, rational discussion and compromise are difficult.

**eukaryote**  A cell whose DNA is contained within a membrane-bounded nucleus.These include such single-celled organisms as yeast and fungi. Cells other than bacteria or blue-green algae are generally eukaryotic. Viruses are not cells and are not eukaryotic.

**F1 hybrid**  The first offspring from a genetic cross between two organisms with different genetic backgrounds. These offspring have an equal contribution of genetic material from each parent. F0 is the parental generation. Crossing two F1 generation organisms produces a second (F2) generation with an equal ratio of parental and hybrid types. Hybrid qualities are rapidly diluted out in the general population if the seeds produced by an F1 crop are replanted by the farmer. Providing only F1 seeds keeps the farmer dependent on the supplier for new seed.

**fermentation**  This term originally referred specifically to the growth of microorganisms utilizing organic compounds as a source of energy in the absence of atmospheric oxygen. It is now used more generally to refer to the controlled culture of microorganisms to produce useful products.

**forensic**  Suitable for courts of law or for public debate.

**gene**  A segment of DNA whose sequence codes for a specific function or protein molecule. The gene is the basic unit of genetic information and can extend for millions of nucleotides in length.

**gene therapy**  A therapeutic approach in molecular medicine where the genetic code is altered in the patient or additional genetic coding is added where there is a deficient gene. The modification can be limited to certain cells of the body and not be transmitted to the next generation (somatic therapy). Alternatively, the genetic information of the reproductive cells can be modified, such that the change is carried through the offspring of the patient into subsequent generations (germline therapy).

**genome**  All nuclear genetic material in an organism, including genes and intervening nucleic acid sequences in all of the chromosomes of an organism.

**germ plasm**  A general term for the living material that controls heredity in a species. It refers to the genetic potential of the species.

**Green Revolution**  A 1960s movement that encouraged intensive agriculture aimed at increasing crop yields through the use of special high yield, hybrid, disease-resistant plant strains and through increased use of fertilizers, herbicides, and irrigation. Its goals were to allow developing nations to become self-sufficient in food production. While these goals were attained, analysis suggests that considerable damage to local ecosystems occurred through overuse of chemical supplements. In addition, residual economic dependency on the fertilizer and herbicides casts a shadow over plans being made to introduce recombinantly modified crops.

**insert**   A familiar name used to refer to the piece of DNA placed or "inserted" into a plasmid used as a genetic engineering tool.

**membrane**   Cellular membranes are composed of two layers of phospholipids, molecules with a phosphate group that faces the water and with a greasy hydrocarbon tail. The tails contact each other forming an oil-like layer to provide a waterproof barrier between the outside and inside of the cell and to isolate the cellular chemistry inside from outside influences. Proteins with special functions to control transport of vital molecules from the environment are embedded in the membrane.

**metabolism**   The cellular chemical processes that provide energy and material for cellular function. A metabolite is any chemical entity involved in these chemical processes.

**microbe**   Microscopically small, single-celled organisms, such as bacteria, algae, and yeast.

**microscope**   An instrument used to make images of small objects. A light microscope uses glass lenses to bend visible light passing through a thin sample to provide magnification of up to 2,000 times. An electron microscope uses magnetic lenses to focus an electron beam, magnifying up to one million-fold.

**monocotyledon**   One of a large family of flowering plants that include the grasses and many cereal grain crops. Monocots have only one leaf in the newly emerged seedling and usually parallel veins in the leaves. They have presented some difficulties in genetic engineering and hence have lagged in development.

**multigenic trait**   A genetic characteristic that is governed by several different genes that contribute to different extents. Many chronic diseases such as heart disease and certain types of behaviors or personality types are multigenically inherited. Drought resistance and cold-hardiness are multigenic traits in plants.

**mitochondrion**   A part of the cell, an organelle, roughly the size of a bacterium, that contains the biochemical machinery to provide the chemical energy needed to run a cell.

**niche**   An ecological term that refers to the specific position in an ecosystem to which its occupant is adapted.

**nucleus**   Compartment of a cell containing the genetic material. In animal and plant cells this region is enclosed by a membrane and contains the chromosomes.

**nuclear transfer**   Replacing the nucleus of one cell and its associated genetic material with the nucleus of another cell. Erroneously dubbed "cloning" by the media, nuclear transfer performed with sheep and with other species produced an identical genetic copy of the nucleus donor, except for cytoplasmic genetic material in the mitochondria. While the transplantation was not true cloning, the feat resurrected the controversy over possible cloning of humans and led to a raft of pending legislation to curb such experimentation.

**nucleotide** The building block of DNA and RNA molecules. Each nucleotide is composed of one of four different chemical "bases"—adenine, cytodine, guanine, and thymine (uracil in RNA)—as well as one ribose sugar and a phosphate group.

**organic** The scientific definition is a chemical compound containing carbon and hydrogen. A popular definition refers to food or products produced without the aid of purified chemicals for nutrients or in processing, often relying on complex biological sources for those materials.

**organelle** A small part of a cell organized to perform a specialized function. The nucleus stores the genetic material, controlling information flow from the DNA to the rest of the cell. The ribosome assembles proteins from information provided by the nucleus through messenger RNA (mRNA). The mitochondrion provides cellular energy and synthesizes needed metabolites.

**pathogen** Any organism capable of causing disease.

**PCR** Acronym for polymerase chain reaction. A molecular biological technique using DNA-replicating enzymes (polymerases) isolated from bacteria living in hot springs or deep ocean hydrothermal vents. This revolutionary method employs cycles at high temperature to rapidly replicate or amplify specific nucleic acid sequences from a complex mixture of sequences. Various modifications allow the technology to perform other reactions using the polymerases to alter target gene sequences.

**plasmid** Pieces of DNA, usually circular, containing genes that can be transferred between microorganisms and into cells and that cause themselves to be copied by the recipient cell. Because they can be handled so easily they are a favorite tool of molecular biologists to store inserted pieces of DNA and to cause the added gene to be activated under experimental control.

**polymer** A chain formed from similar types of building blocks. Typical biological building blocks and their common polymers are amino acids (proteins), carbohydrates (cellulose or starch), and nucleotides (DNA and RNA). Nonbiologic polymers include tars, waxes, polyesters, and polyamides (nylon, plastics), although genetic engineering is allowing plants to carry out some of these syntheses.

**prokaryote** A cell such as a bacterium whose DNA is not contained within a membrane-bound nucleus. The DNA usually is found in a single long, circular strand. Compare *eukaryote*.

**protein** An unbranched chain or polymer made up of units called amino acids. There are twenty types of amino acids. Proteins provide structure and catalytic activity for cells to synthesize and organize the rest of their components.

**recessive trait**   A single-gene genetic characteristic that is only evident when both chromosomes of a particular pair carry that particular DNA sequence. Among genes controlling human eye color, the gene for blue eyes (in adults) is a recessive gene. Compare with dominant gene.

**recombinant DNA**   The manipulation of DNA and genes in new combinations. While this process also takes place in nature, this term is normally used to refer to human-mediated rearrangements.

**replication**   The copying of the DNA sequence of the genome.

**restriction enzyme**   An enzyme (endonuclease) capable of recognizing a specific DNA sequence of four or more base pairs and cleaving the double-stranded helix at that sequence. Evolved by bacteria to prevent incorporation of foreign DNA into their genomes in a sort of "genetic immune system," these enzymes are isolated by molecular biologists and used to cut and paste DNA segments in the desired order. The ability to manipulate DNA combinations at will was the key to the recombinant DNA revolution and genetic engineering.

**reverse transcriptase**   An enzyme produced by viruses whose genetic material is RNA rather than the usual DNA. It possesses the unusual property of copying RNA sequences into DNA (an RNA-dependent DNA polymerase). This property has been exploited by molecular biologists who use the enzyme to copy messenger RNA into a complementary DNA sequence, which can then be replicated and engineered in the standard way in bacteria and other organisms.

**ribosome**   A part of the cell (an organelle) that organizes the connecting together of amino acids to produce a protein.

**species**   A group of like organisms classified together. Individuals within a species can interbreed. Different species cannot interbreed to produce fertile offspring.

**strain**   A variant within a species. Intraspecific strains can interbreed.

**transcription factor**   A protein that regulates the activity of genes. It binds to a specific DNA sequence in the regulatory part of a gene and organizes several proteins and the enzyme needed to synthesize messenger RNA.

**transgene**   A gene from another species of organism.

**transgenic organism**   An organism containing genetic material from another species. A mouse engineered to contain the human hemoglobin gene is a transgenic animal. It contains artificial combinations of genes that would not have occurred with any appreciable frequency in the natural environment.

**tumorigenic**   Capable of causing a cell to divide out of control and become cancerous.

**vaccine**   A preparation of killed or weakened pathological organisms or parts of those organisms administered to induce immunity against that pathogen.

**vector**   A DNA carrier used to replicate an inserted DNA sequence. These vehicles were derived from naturally occuring self-replicating DNA circles found in bacteria and in uekaryotic cells. They can also direct synthesis of mRNA from the inserted DNA sequence.

**virus**   A small protein and lipid particle containing DNA or RNA sequences that can penetrate cells and direct their own replication. Parasitic, lacking metabolic functions, and unable to replicate themselves without assistance, viruses are generally not considered to be alive.

**xenograft**   Transplantation of cells or tissue from one species into another, such as replacement of a human heart with a baboon heart.

# Abbreviations and Acronyms

**AAV**   Adeno-associated virus

**ADA**   Adenosine deaminase, an enzyme involved in nucleotide metabolism

**AIDS**   Acquired immunodeficiency syndrome

**ALS**   Amyelotrophic lateral sclerosis, a neurodegenerative disease also known as Lou Gehrig disease

**APHIS**   Animal and Plant Health Inspection Service

**BSCC**   Biological Science Coordinating Committee

**Bt**   *Bacillus thurengensis* toxin

**CEO**   Chief executive officer

**CF**   Cystic fibrosis

**CRADA**   Cooperative Research and Development Agreement

**DOE**   Department of Energy

**ELSI**   Ethical, Legal, and Societal Implications—Task Force of the Human Genome Project

**EPA**   Environmental Protection Agency

**EST**   Expressed sequence tag

**FASEB**  Federation of American Societies for Experimental Biology

**FBI**  Federal Bureau of Investigation

**FDA**  Food and Drug Administration

**FSIS**  Food Safety and Inspection Service

**GATT**  General Agreement on Tariffs and Trade

**G-CSF**  Granulocyte stimulating factor

**HD**  Huntington's disease, a neurodegenerative disease affecting muscular coordination

**HGP**  Human Genome Project

**HGS**  Human genome sciences

**HIV**  Human immunodeficiency virus

**HMO**  Health maintenance organization

**IND**  Investigational new drug application

**NIH**  National Institutes of Health

**NSF**  National Science Foundation

**OTA**  Office of Technology Assessment

**PCB**  Polychlorinated biphenyl (a complex environmentally stable organic molecule)

**PCR**  Polymerase chain reaction, a technique used to replicate or amplify specific nucleic acid sequences

**PMN**  Premanufacture notification

**RAC**  Recombinant DNA Molecule Program Advisory Committee

**RFLP**  Restriction fragment length polymorphism (pronounced "riff-lip")

**S&E**  U.S. Department of Agriculture, Science, and Education

**TIGR**  The Institute for Genomic Research

**U.S.C.**  United States Code (legal code)

**USDA**  United States Department of Agriculture

# Index